Customer Experience Management Rebooted

Steven Walden

Customer Experience Management Rebooted

Are you an Experience brand
or an Efficiency brand?

palgrave
macmillan

Steven Walden
TeleTech Consulting
London, UK

ISBN 978-1-349-94904-5   ISBN 978-1-349-94905-2   (eBook)
DOI 10.1057/978-1-349-94905-2

Library of Congress Control Number: 2016958038

© The Editor(s) (if applicable) and The Author(s) 2017
The author(s) has/have asserted their right(s) to be identified as the author(s) of this work in accordance with the Copyright, Designs and Patents Act 1988.
This work is subject to copyright. All rights are solely and exclusively licensed by the Publisher, whether the whole or part of the material is concerned, specifically the rights of translation, reprinting, reuse of illustrations, recitation, broadcasting, reproduction on microfilms or in any other physical way, and transmission or information storage and retrieval, electronic adaptation, computer software, or by similar or dissimilar methodology now known or hereafter developed.
The use of general descriptive names, registered names, trademarks, service marks, etc. in this publication does not imply, even in the absence of a specific statement, that such names are exempt from the relevant protective laws and regulations and therefore free for general use.
The publisher, the authors and the editors are safe to assume that the advice and information in this book are believed to be true and accurate at the date of publication. Neither the publisher nor the authors or the editors give a warranty, express or implied, with respect to the material contained herein or for any errors or omissions that may have been made. The publisher remains neutral with regard to jurisdictional claims in published maps and institutional affiliations.

Cover image © erhui1979/Getty

Printed on acid-free paper

This Palgrave Macmillan imprint is published by Springer Nature
The registered company is Macmillan Publishers Ltd.
The registered company address is: The Campus, 4 Crinan Street, London, N1 9XW, United Kingdom

Do you think Apple, Zappos, Geek Squad and Amazon achieved 'Experience' brand status by just focusing on 'Efficiency'?

Do you think companies only create customer value by being the best at fixing problems?

Create experiences that motivate not just manage 'experiences' that don't?

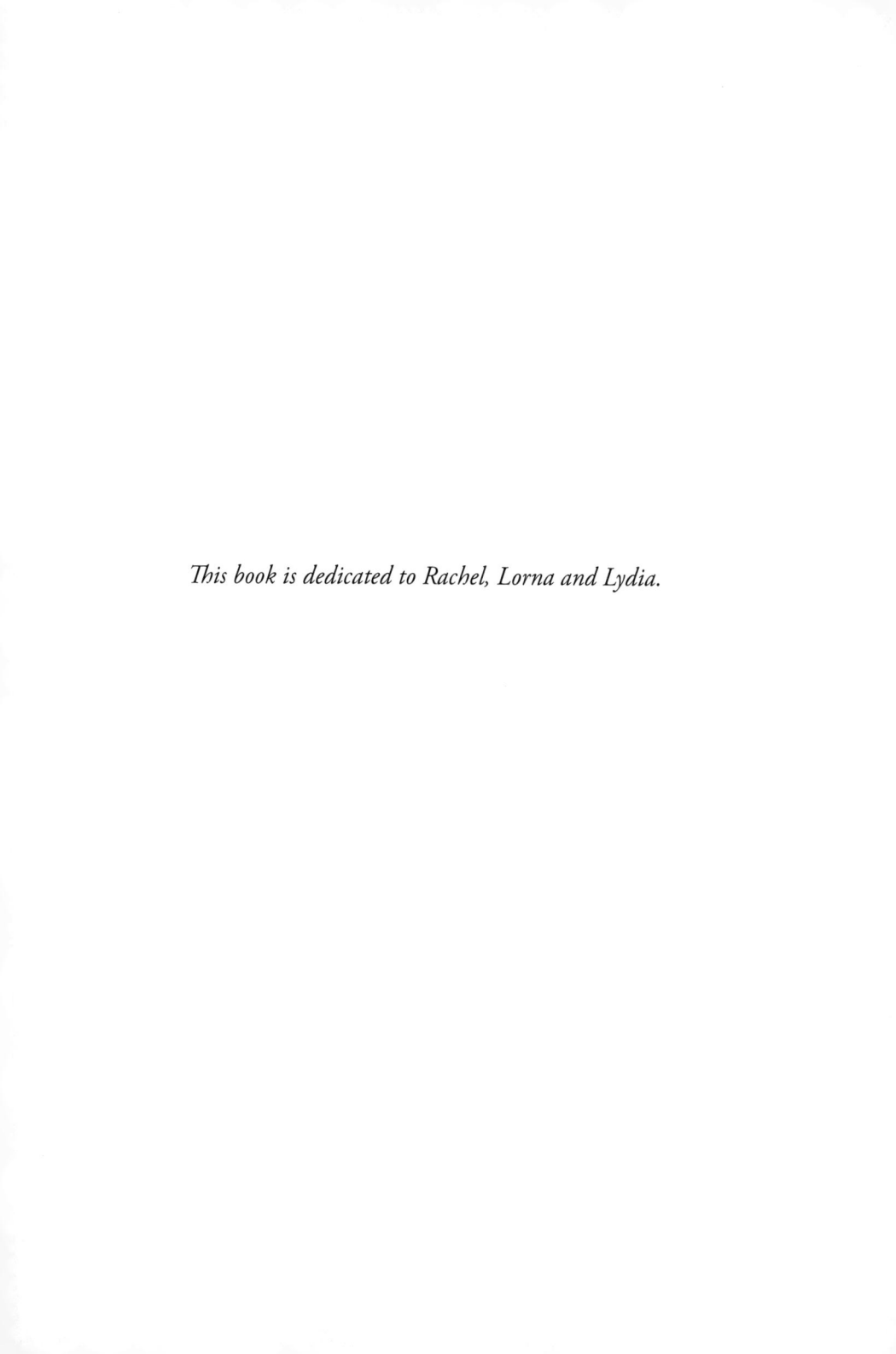

This book is dedicated to Rachel, Lorna and Lydia.

Preface

There is no technology in this book.

Technology is an enabler of the customer experience (CX); it is not 'the experience'.

For instance, no customer buys 'the website'. They buy the benefits of using the website; to book that flight, to browse on iTunes[1]. Hence, it is these frequently underlying and intangible customer motivations that we have to uncover and create if we are to 'do' customer experience. Otherwise we risk investing our money in tools, processes and methodologies that focus only on achieving efficiency for the firm rather than on memorable customer benefit[2].

Which is, of course, exactly what we see in today's market, deluged as it is by so-called customer experience management (CEM) solutions!

Hence, the mission of this book is to *reboot* our understanding of customer experience; to make us realise that we need to put the customer back into customer experience management (CEM). After all, when the unit of measurement is the human mind, the aim of CEM as a strategy must be to create 'Experience' brands.

For instance, consider how Apple, Zappos, Starbucks, Geek Squad, Giff-Gaff, Overbury, Cerritos Library, LUSH, Hotel Chocolat, Stew Leonard's, Disney and Amazon became so successful. They did not create competitive differentiation by focusing on efficiency or even product and service delivery alone but by creating 'experiences' in the mind of the consumer. And not any old experience, but authentic personal and memorable ones that are 'meaningfully different' and connect with customer value creation.

Experiences that are 'memorably' effortless, seamless as well as experiential! While of course having one eye to efficiency – but efficiency without contradiction.

And they were able to do this, because they knew what it is like to be a customer. They had empathy.

Unfortunately, this message now seems to have been lost. Which surprises me for while efficiency is important to the 'experience the customer has', there is no point in using a term like customer experience unless it means something else![3]

And remember, even if we do go down the path of confusing 'customer experience' with firm 'efficiency' gains, we cannot claim even these things for 'experience' if we fail to show in their execution how they have improved the 'subjective' experience the customer has!

Hence, I argue that in order to do customer experience we don't need to be ever more obsessed with efficiency. In most cases, customers will, rightly, assume that these things should be a given, they are seen as simple basic competence and do not move us as consumers.

Customer experience is not about making things so efficient that customers don't notice them.

Even worse, since creating experience is a design principle involving trial and test, the actions of constraining command and control efficiency processes can limit our capability to create experience if taken too far.

As an example, think about how the animation company Pixar makes films such as *Toy Story* and *Monsters Inc*. Typically it takes over 100,000 storyboards to make one film of 12,000! At one level that represents an immense amount of waste, but without it there would be no great film. Now imagine what would happen if Pixar only allowed their teams to create 12,000 storyboards per film!

Recasting Our View

Hence, the current dominant approach to CEM is flawed.

It is also not right to think physical use (poorly conceived user experience) is equivalent to psychological engagement (customer experience). By following such an approach we lose the intent of customer experience: to differentiate based on the provision of a *memorable* experience.

Therefore, we must recast our view of customer experience and accept that it is concerned with creating value from how customers think, feel and behave.

But of course, many tools vendors will not agree with this, but all I have seen over the last 10 years from their approach is a CEM market that has *lost its marbles!*

For instance, companies now use the term 'experience' as an over-brand for anything that vaguely touches 'the customer' or they race around summing up touchpoints to produce a meaningless number, desperately concerned with hygiene.

In my view, this represents the sale of software based on a ridiculous belief that consumers respond to 'experience' like a machine: a viewpoint at best touchingly naïve, at worst coming close to selling snake oil. It also leads to a deep disconnect between the C-suite, looking at their dashboards and what actually happens on the ground, in the minds of our consumers and employees.

Yet you also have to accept that maybe this focus is creating a platform for something else.

All this effort may well end up leading us back to experience, based on the design of disruptive personalised journeys and putting real power in the hands of the consumer not the firm.

United Kingdom   Steven Walden

Notes

1. A cognitive assessment of experience is about intangible benefits as well as tangible. That is why a focus just on the tangible such as 'website', 'network performance' or 'booking' misses the point that customers see the world qualitatively, influenced by their subjective senses, based on 'feel' together with reason. Examples of this include their sense of trust, how satisfied they are, the 'feel' of the brand and so forth.
2. An objective-only definition will focus investments on technology, which is a considerable cost in terms of asset base if not connected to customer value. In this way, it may constrain opportunity-seeking behaviour, the ability to look for frequently small effective changes in the experience that mean something to the consumer. So a smooth tech platform that works, while being very important, would not on its own deliver a LUSH or Metro Bank experience. Nor would it lead to better people engagement.

    A technology-focused environment also risks creating an obsessive focus on risk management and hygiene factors with 'operators' confusing machine data with 'that must be how the customer thinks.'

    Hence, we lose the capability to focus on intangible benefits, the very thing consumers buy. Tangibility is only important as an enabler of intangible benefits.
3. This is what happens to many marketing concepts. You can't scale an idea or expertise, but you can scale software. So, if we take the experience the customer has, then feed it through the mincer, TQMing it, breaking it down into components and in the process miss out on how cognition really works, we can measure and monitor it. And voila we have a practice. This has the unfortunate effect of giving us false hope while at the same time failing to account for experience losses.

Contents

1 The Squonk   1

Part I   Understand   7

2 Right Understanding   9

3 Right Commercial Principles   33

Part II   Data   73

4 Right Data   75

5 Some Key Things That Make Subjective Data Different from Objective   83

6 The Subjective Data Line   97

7	Customer Experience Is Complex	107
Part III	**Customer Experience Research**	**121**
8	Traditional Surveys Are Efficiency Surveys	123
9	Best Practice CX Research Methods	129
Part IV	**Emotions and the Customer Experience**	**151**
10	The Value of Emotions	153
Part V	**Mindset**	**185**
11	Right Mindset	187
Part VI	**Not Do**	**205**
12	Customer Experience Bad	207
Part VII	**And Finally**	**217**
13	Interconnectedness	219
Footnotes		241
Index		249

About the Author

Steven Walden has taken a number of senior roles in CX.

He is currently Director of CX at TeleTech Consulting, a world-leading customer experience firm comprising leading brands in culture and mindset, analytics, loyalty, service excellence, sales transformation, as well as technology and call centre operations. His role encompasses consulting and thought-leadership through the leading CX professionals network: CX in Action.

He has also been Director of Customer Experience at Ericsson and prior to this he spent over 8 years as Head of Consulting and Research at boutique customer experience research and consultancy firm Beyond Philosophy.

In Ericsson, he designed and managed their Experience Management Centre (EMC) and CEM Partnership programmes; rolling out global best practice in social media analytics, CX maturity assessment and customer experience measurement using, in particular, story based metrics. Agency side, his insights, techniques and strategic approaches have been used by many FTSE and Fortune 500 firms in their customer experience programmes.

In addition, his CX techniques won the UK CEM Awards 2013 (insights category 'love the customer') while the methodologies he created for emotional measurement and design have been cited as best practice by Forrester.

Steven is a keen advocate for integrating complexity science with customer experience: especially in the area of measurement using stories (he ran the first NPS study using stories as the measure) and in its approach to culture and innovation.

In addition, Steven regularly speaks at conferences, authors books on customer experience, conducts thought-leadership research, much of which has been published in leading business journals, and writes/blogs on CX.

List of Figures

Fig. 2.1	DEM approach to understanding customer experience	18
Fig. 3.1	Subjective asset management in a service	39
Fig. 3.2	SAM Tier 1. Service and product efficiency	42
Fig. 3.3	SAM Tier 2. Service and product excellence	49
Fig. 3.4	SAM Tier 3. Product and service drives	53
Fig. 5.1	Subjective experiences are fuzzy composites	87
Fig. 5.2	Figure–ground in CX	90
Fig. 5.3	More of objective dashboard	92
Fig. 5.4	Change meaning subjective dashboard	93
Fig. 6.1	Perception data curve	98
Fig. 6.2	Managing resilience	102
Fig. 6.3	The blue dot effect	103
Fig. 7.1	Complicated effects	110
Fig. 7.2	Complicated NPS	110
Fig. 8.1	Traditional view and customer experience view	124
Fig. 10.1	Emotix study	159
Fig. 10.2	Emotion framework	179

List of Tables

Table 2.1	Why text algorithms don't measure emotional drives in entirety	16
Table 3.1	Customer account and firm account	34
Table 3.2	Summary of personal gains from excellence	51
Table 3.3	Summary of customer experience approaches	60
Table 3.4	Quality questions	70
Table 12.1	Myopic inside–out view	214

1

The Squonk

Experience Versus Efficiency

Just because someone says it's so, doesn't mean it is. *Don't Believe the Truth* as the old Oasis album title goes. Kind of like if I say 'customer experience' and then immediately start to talk about 'efficiency', I mean are we talking about the same thing? Sure as a service paradigm 'efficiency' (by dictionary definition: the least waste of time and effort) is fundamental but 'the experience the customer has' is a cognitive assessment that must mean more than service otherwise why use the term!

I have been to many CX conferences; listened to hundreds of executives and worked with tens if not hundreds of firms and what strikes me most is this contradiction at the heart of CX. For on the one hand we want to engage the customer, be memorable, build loyalty, and create a degree of emotional commitment like Amazon, Geek Squad, Disney and Lush. Yet at the same time we believe we can get there through efficiency, fixing breakages, summing all touchpoints and leaving it at that.

For instance, at one recent conference in London, I listened to a panel of CX directors talk about the omni-channel experience and how they were committed to making it so consistent and easy that no customer would even notice it's there. It just works; it's service delivery.

© The Author(s) 2017
S. Walden, *Customer Experience Management Rebooted*,
DOI 10.1057/978-1-349-94905-2_1

Contrast this to the next discussion.

Here the speakers included a medical devices vendor that had created personal and memorable experiences using animation to help children feel relaxed when they go into an FMRI scanner. Then there was the US payments provider who gained market share by delivering a personal feel to their money transfer services, achieved through the use of local community Spanish speakers in Mexico to engage with newly arrived Hispanic migrants. Making them feel at home when they transfer their money! These firms had thought beyond service delivery.

Are you still sure your efficient tech platform alone is delivering an 'experience'?

For me it's clear. Efficiency is about, well, making things efficient. It is an 'objective' characteristic. The doors work, I can get on the website, I receive a signal on my handset, my pin number is accepted, the floors are clean, the lights come on, the technology works and so on and so forth.

Efficiency is about how well resources are allocated and hence how well potential operational failures are controlled, the workability of things – its efficient it doesn't break – not how well the customer experience is being met.

So although efficiency is a critical enabler of CX, it is still about the transactional, the hygienic and the forgettable 'experience' (there is no impact on most cognitive assessments). However, it has one killer application – it's measurable – literally the sum of all touchpoints. Hence, it is web click through rates, web browsing and video download speeds. And with the Internet of Things, it will soon get a lot worse; expect plenty of measurements but no customer included!

I would contend that although such analytics are necessary, they only become CX analytics if the customer is included.

But don't get me wrong. I am not antiefficiency. It is fundamental. It's just that when we talk about it we tend to mean avoiding mistakes or cutting costs. And however noble that is as a cause, this is not customer experience. It is service assurance reborn, a philosophy of zero defects circa 1990, a belief that the best experience is 'no experience.'

It's nothing new.

In fact, it can be the opposite of customer experience.

That's because customer experience is not an objective but a subjective thing, the whole point being to go beyond commoditisation and 'efficiency'. To focus our business not on our own processes but the process that goes on in the mind of the customer. To see as they see, to grasp the nettle of customer reaction and build the customer view into our business model and create 'experiences that matter'; experiences that are *personal and memorable,* experiences that motivate and drive consumers; experiences that are 'meaningful', 'different', 'noticeably easy and seamless'.

It doesn't matter if this means it is only relevant in a few circumstances. What matters is that its meaning is clear.

So how has this confusion arisen?

The Bastardisation of Customer Experience

It seems that business is full of examples of concepts that become bastardised; once a term achieves popularity, every man and his dog will use it to sell their products and services. Hence, customer experience (and its management) may well have started off talking about creating personal and memorable 'experiences' through looking beyond service attributes but ultimately this subjective approach was ignored in favour of an objective one: as in how we use experience to mean 'everything I experience'.[1]

After all, how can you sell *personal and memorable* experience if it depends on an idea and its execution? And what are these experience attributes that go beyond service, product and price anyway?

For instance, consider the Casa del Agua in Mexico, showcased at the Forrester CX Forum 2015. The idea here is to sell bottled rainwater after it has been passed over rocks carved with emotionally expressive words. It's a fantastic concept, breaks the mould and sells well. It has created a new market space! It has high emotional value! But it is also specific to the circumstance and requires human creativity to enact.

Now imagine talking to CFOs and using this as an example of experience. If they are not open to defining a new market space through emotional value or investing in trial and test, then it will not even get off the ground. For them, it sounds far better to monetise 'experience' by selling or investing in a box to measure and monitor everything!

In this way, ROI is achieved for the vendor; *'experience'* for the customer is not.

And this kind of bastardisation is nothing new. Take, for instance, another concept: customer relationship management (CRM). This started off talking about building a *relationship* with the customer until the term 'relationship' became defined as the ability to cross-sell and up-sell off a database.[2]

ROI was achieved for the vendor; *relationship* with the customer was not.

The Qualitative Difference

So let's be clear, to use customer experience as a source of value differentiation means we must take account of its qualitative, phenomenological nature. We must define what customer experience as a 'cognitive assessment' means: where it differs from price, product and service quality and how it relates to customer value creation (e.g., the personal and memorable bit). In fact, in an environment where over-inflated data promises are spreading viruses such as taking such an approach to understanding customers has never mattered more.

> Company, can you help us with our CEM programmes?
> Vendor: Measure and monitor everything; yes we can do that. It will of course mean building rocket science predictive models and putting in place extensive automated software systems. Here's our ROI proof point against NPS using Advanced Regression. And of course we consider qualitative opinion; we take in a mass big data feed from social media and text verbatims applying a machine-based algorithm that spits out in realtime correlates to your key KPIs. Have a look at our nice shiny dashboard, architecture chart and proof of concept. That'll be £1 million.

So, tell me what exactly is the customers *experience* after you put this system in place? Do you really believe that such an approach in toto picks up the real nature of experience?[3]

Analytics is the first fundamental step in any CX programme (Know where you are! Predict where you want to be), so we should really do a better job of tying analytics to experience to get the result we need.

Which is why, perhaps when it comes to customers and their experience, by seeking to control it we are actually degrading it by missing out on the human dimension. Maybe the obsessive management of risk is the greatest risk of all? Maybe when it comes to the experience the customer has we need better data not big data and ensure that our AI processes are there to augment not replace human understanding, the relationship building process and how we handle emotion.

A Squonk of a Thing

Hence, it feels to me that the time is right to reappraise customer experience and provide support to the many businesses who want to move away from competing just on price or operational efficiency to competing on customer experience. This is why I seek to clarify the confusion surrounding CX and set out a series of 'right' approaches, paying particular attention to the problem of how to measure it. While all the time being clear that we must separate out efficiency from experience lest we conflate the two and end up with a suboptimal result.

It is also clear to me that hunting the meaning of customer experience is like hunting the Squonk.

> The legend holds that the creature's skin is ill-fitting, being covered with warts and other blemishes and that, because it is ashamed of its appearance, it hides from plain sight and spends much of its time weeping. Hunters who have attempted to catch squonks have found that the creature is capable of evading capture by dissolving completely into a pool of tears and bubbles when cornered....
> Source: Wikipedia

No one can claim he or she has captured the CX or CEM Squonk but if we can at least identify what it might look like then we might have a chance of 'doing customer experience'.

And if your answer after all this is still 'buy this tech box', 'personalise that omni-channel experience' then fantastic, you have tied your actions to improvements in customer value, how customers think and feel (which doesn't necessarily mean NPS!) and you are doing customer experience.

My final message is this. If you don't believe in customer experience as a differentiator, don't do it. If you don't believe in making money out of the 'experience the customer has', then don't do it. Opex reduction, process efficiency, cost cutting and sales growth, never mind the 'experience,' are all fine and well-established principles. I have no problem with them.

Just don't pretend you're doing 'customer experience'.

At the end of the day CX cannot claim all and every benefit and nor should it. Sometimes, frequently even, there is a place for other ways of doing business, but without a clear idea of what experience is and its limitations we will never know what it means and where to apply it. And that way will only lead to its death.

Notes

1. Customer experience has become a general term for: (1) service design, (2) seeing things through the customer's eyes and (3) serving up any old thing as 'selling experience'. Apart from the last item, this is not a bad thing. But it does mean it holds little difference in meaning from say service and customer research! Hence, service design should really be experience design.

   In addition, the complexity of digitalisation and multichannel integration has meant that the words 'customer experience' are just useful terms with which to talk about user experience. This is all fine, but an over-brand is not enough, even if it does give a sense of mystique (not quite definable, slippery, can mean what you want, magical!). We need to consider the experience economy definition and how we can create an experience not manage one.

2. I am aware that relationship marketing is the origin here and that there is a clear ROI of CRM based on software. However, I stand by my comment: if you use a qualitative term like relationship then you need to demonstrate a qualitative benefit!

3. If you take a technology enabler to be 'the customer experience', you risk assuming that when the machine KPIs work well then the customer must be happy! This is not necessarily the case if the customer thinks of the 'enablement' as hygienic. And certainly not the case because no customer buys the tech enablement; they buy the benefit that's derived.

   Likewise, where do you stop your technology enablement as an experience definition?

   A back-office document bill printer has a critical role to play in the experience of electricity and gas buying. Is this now a customer experience solution?

Part I

Understand

In this section, we look at how false notions of customer experience have led to a focus on creating the Efficiency brand. To counter this we propose a proper definition of customer experience and its implications for what we do. We start with this section because everything flows from 'understanding'.

2

Right Understanding

What Is Customer Experience?

The concept of 'customer experience' has always been with us. You only have to think of the 1992 film, *Home Alone 2*, starring Macaulay Culkin. About 30 minutes into the film we hear the New York Plaza Hotel advertised as 'New York's most exciting hotel experience.'

Two years later Carbone and Haeckel raised the profile of experience further when they spoke of it as 'the takeaway impression formed by people's encounters with products, services, and businesses – a perception produced when humans consolidate sensory information.'

However, from a commercial as opposed to academic point of view, it was only in the late 1990s through the work of Pine and Gilmour in their *Harvard Business Review* article 'Welcome to the Experience Economy' (1998) and book *The Experience Economy* that the movement for customer experiences really kicked off. Especially because this formalised customer experience as an economic principle in its own right 'as distinct from services as services are from goods' (Joe Pine).

And the implications of that are: we must put how customers think and feel at the centre of our business activities and understand how this differs from notions of price, product and service delivery.

However, since the late 2000s things have become considerably murkier. Indeed by 2010, IT analyst house Gartner had found over 14 different definitions of customer experience, a figure that stands considerably higher today.

So what are the origins of this confusion?

Well, I believe problems have arisen with the mass entry of CRM vendors and consultancies into CX which together with the explosion in complex digital, IT and UX projects has confused the market. Turning the phrase into, ironically, a service assurance concept circa 1990! In effect a useful foil to simply say, think about the practical and objective experience when customers use your brand. Aim for zero defects, something customers don't even notice! Or if they do, it hardly motivates at all; it is just the ticket to the game, 'a thin gruel' of a definition of experience and certainly one that doesn't create an Experience brand. In other words 'No customer psychology included!'

Or how about the common practice amongst CX practitioners of thinking experience is about 'everything'! As if customers are desperately concerned with every detail of their engagement with you and react accordingly. As if 'the takeaway' impression is a summation of rather than a function of all that is 'experienced' before!

But for me, these reworking's miss the point; they reconceptualise experience as efficiency by treating subjective experience as objective! Focused on achieving the least waste of time and effort for the company in delivery – hence making things work – but the customer be damned!

Which means of course that customer experience is now in danger of losing its meaning: after all managing for the 'nonexperience' is not a customer experience concept and 'Come to the New York Plaza Hotel because the carpets are clean and things work,' is not a great slogan.

Which is why when companies consider the question, 'What is customer experience', how they answer it holds important implications for the validity of their customer experience programmes.

Think of it this way.

If you try to measure and manage something but you are unsure what that something is you risk measuring and doing the wrong thing. Rather as if if you were interested in losing weight but measured your height. In these circumstances, a diet will never be on the agenda.

Likewise with customer experience: if it ends up meaning something other than what is was intended for, say, process efficiency or anything you want, well don't be surprised if you end up with a suboptimal CX result.

Remember something that means anything to anyone really means nothing at all.

Hence, if we can define customer experience, then we will know how it delivers value, how to measure it and what to do about it. If we can't or just use it as a rebrand then it is just so much marketing fluff.

And yet, you have to ask yourself, what's the problem?

The Experience the Customer Has

What is customer experience? It's the experience the customer has, in other words their subjective experience. How they think, feel and behave. It's not difficult. There is a whole industry called market research devoted to understanding it.

But don't just take my word for it, Meyer and Schwager from their 2007 *Harvard Business Review* article write: 'Customer experience is the internal and subjective response customers have to any direct or indirect contact with a company' (Meyer and Schwager, HBR, 2007).

Or how about Joe Pine and James Gilmour, the original and key authorities on customer experience? For them customer experience is about the *personal and memorable* moments, the value of time well spent.

For Meyer and Schwager and Pine and Gilmour customer experience is about subjectivity, which by definition means it is qualitative and phenomenological, the unit of measurement, the critical point, being the memory of it all.

Hence, in this way customer experience has extended our understanding of the customer beyond notions of product, service and price. To quote Olaf Hermans:

> We have to give credit to experience. It is not so much about emotions as the phenomenological aspects – considering the holistic. Hence, it said goodbye to SERVQUAL, the consideration that customers are judgers and

instead, focused on how things come to customers, how things are cognitively received and felt. This is more grounded and is a better way to think than the critical performance dimensions.

What it comprehensively is not, then, is experience management without the customer!

Redefining Experience as Drives

Yet, I believe this definition of customer experience by Meyer and Schwager is not enough because not everything we subjectively experience matters. And surely for an 'experience' to be 'an experience' it must matter, rather than be some subjective 'event' that flops in and out of consciousness or even subconsciousness.

So how can we craft a definition of subjective experience that differentiates between those billions of experiences we receive each day that are unimportant from those few that are? Well, I would contest that the valuable ones are those that drive us. Let me explain by taking an analogy from emotion theory:

> Imagine a football match. You have invited your friend along to see the game in which your favourite team is playing. Your friend doesn't much like football but is coming along anyway. Halfway through the game your team scores a goal. What's your emotional reaction? Happiness, of course. What's your friend's emotional reaction? Well, he might be happy for you, but couldn't really care less about the game.

Here, the same game was being watched but the reaction was very different, because the *drive*, the impulse to react was different.

We also see this point coming through in a famous experiment conducted by Hasdorf in the 1950s. Here after a particularly rough game of football between two US colleges, Dartmouth and Princeton, students of each college were taken aside to view a cinematic replay of the event. Even though both sets of students viewed exactly the same footage each side blamed the other for the bad behaviour and foul play.

Once again their drives were different so their personal and memorable experience was different.

So when we talk about customer experience for me what we really mean is:

> Those subjective experiences that lie in memory or are experienced in the moment, that influence and are influenced by our drives which in turn lead to behaviour.

For me, the unit of measurement is *what drives the customer*! Memory by contrast is only one aspect of consumer response and without a *drive* risks being less durable: a memory can be just a memory; a *drive* encompasses memory but is more connected to value.

Customer Drives

So if drives are so important, what are these? Well, for me *drives* encompass the expressed and unexpressed needs of customers (their goals and subgoals), which of course involves a multitude of influences: sometimes rational, sometimes emotional, sometimes about affect, sometimes a conditioned response, sometimes subconscious and so it goes on.

However, the reason I don't use the term 'needs' is that this tends to focus the business on a simple rational view of the customer. Drives feels to me to focus things more broadly (Source: Dr Simon Moore).

So although a traditional view holds that needs have a basic and direct root-cause relationship to behaviour as in 'I need that car' or 'I need these grocery items', drives encompasses needs AND a broader, more holistic range of indirect and in the moment experiences 'important to me'. For instance, drives include:

- How a trip advisor comment on a hotel influences whether I want to go there
- The sensory in the moment appeal of a brand logo
- The responsiveness of an NHS hospital to care and availability of blankets

- The warm empathy of the Marks and Spencer's call centre rep
- Whether Etihad Airlines has a well-stocked library of music on its flight

What also drives me then are these small in the moment experiences that sometimes indirectly influence how I feel about doing business with you even though on their own they may not be root-cause effects on behaviour as one might find with a 10% drop in price leading to a 1% rise in sales.[1]

So, in a hospital, availability of blankets may not have an ROI attached to it, but it is part and parcel of an overall sense of care that you should look to build because it is an important correlate to satisfaction, NPS and whether you would use the hospital again.[2]

Likewise, on that flight, the price and seating may be undifferentiated from the competition. But what I remember and what does differentiate is how well or badly the music library is stocked! Even if I do say, I bought on price!

The impact of intangible drives can also be seen on the BBC programme *The Fixer*. In one episode, Alex Polizzi went to a children's play area called The Big Space. Here she managed to raise sales by 5% per head through 'theming' the experience alongside a focus on cleanliness and better food quality. Of course 'theming' does not have an ROI attached to it, but does when added to other elements of the experience.

In this way we can see that understanding drives is important for effective experience design; as long as we understand 'what matters' which most certainly does not mean everything and frequently means an intangible aspect of the experience sometimes nonconsciously and fleetingly expressed.

Missing the Drive

The irony is firms fail to focus on what drives the customer because they are mostly wedded to an engineering model of experience that reduces our understanding of drives and by consequence emotion.

For instance, the enterprise feedback management vendors are certainly interested in emotional drives but only to sell their service lines, while at the same time constraining the debate on experience and creating emotional *value*.

So now with EFM we can fire off emails to capture how you *feel* close to the moment. We can then hunt down root cause by using artificial intelligence and operationally respond to the issue at hand. We can then measure the result through declines in negative emotions (sentiment score) and link this to churn and complaints data.

Sounds great, and I like it! Except, this is mostly an efficiency not an experience methodology; think of it this way.

Imagine I have just been to my local fish and chip shop. I leave and then get a text message on my handset asking me about the performance. The conversation might go something like this:

Machine: How was your experience at fishnchicken today?
Customer: Yeah it was alright, same as always, 8 out of 10 (damn text message again).
Machine: Chunk chunk chunk, a passive; feed more questions to understand why, what are the root-cause drivers; how can we turn him into a promoter?
Customer: Got to get into the house before these chips go cold.
Machine: Why did you say it was alright?
Customer: Because it was alright (I really like fishnchicken I gave it an 8 out of 10 because it's good; it's just fish and chips; I always go there).
Machine: Can you let us know a little bit more?
Customer: The fish and chips are good value for money and taste nice.
Machine: Hmmm…functional price and value are the main criteria of the experience and root-cause drivers.

End of conversation.

So what can we conclude? Well through our operational root-cause and binary focus we have managed to miss what also drives the consumer! For instance consider even in a flabby positive comment like this we can see the following issues as outlined in Table 2.1 (and please note this is not a critique just an identification of differences).

And anyway, a piece of text is not the same as a person, however fancy your algorithm.

So, this kind of thing is not wrong it's just limiting! Of course there is a practical benefit, as there always is. But my point is, the practical benefits

Table 2.1 Why text algorithms don't measure emotional drives in entirety

Issues with text algorithms	Commentary
A recommendation score of 8 out of 10 is fine	It's consistent with the context; it's a great score! The NPS rule turn passives to promoters is not a rule
I may be gifting a score not because the score drives spend but because spend drives the score!	It's use and tenure that really matter
My text sentiment is not the same as how I feel about you most of the time	You haven't measured feelings. I like you, but I don't feel great all the time and sometimes I feel bad
You haven't measured my overall sense of the experience	Played out more in the decision-moment prior to me leaving the house! My drives
I am affected by things outside the functional, such as my situation (must get home!); ego, social identity and personality	So you really have not thought about the full experience anyway
I always answer a direct question directly	It's about the functional but there are other experiences indirectly impactful I have not reported such as the fact I visit the charity shop at the same time; it's convenient, the ambience is good
Maybe the functional is good enough	Product quality has a curvilinear relationship to spend, something I investigated in my emotion work
I only know what I know and the shop could try selling other things	Curry chips anyone! Or perhaps you could compare yourself to exemplar experiences, the 'to be'
This tells me nothing about change effects	So if I do invent curry chips, what next? How will it affect other things, how will it be executed
I didn't think anything much until you asked	It's a habit
Liking or disliking doesn't always relate to a root-cause with an algorithmic relationship to spend	Subtleties exist: some I don't even know about!
I may not know what I want and what could be	Customers like me frequently sacrifice to blandness
You have missed out on nonverbal, nonconscious impacts	Critical to our understanding of drives, for example, the fact I know the staff, helps to embed a positive relationship
You have missed how things change	Next week, I've had enough of fish and chips

have an impractical result – we fail to build the Experience brand! Which I thought was your point.

But don't get me wrong. We can find some great things by accessing text; such as the story of how one vendor collated fairly low in the moment results for an airline route to Las Vegas. On further investigation through AI they found that this was because customers were in a different party mood and more responsive to an alternative experience; or how about how Homebase optimising services from text dashboards again delving into comment and root cause.

But I have to say there is another sting in the tail of all this. If we over-constrain things by only obsessing about what customers say we miss out on the employee experience.

For instance, as an employee I got one thing wrong and now you're beating me up! Well I'm not going to push the boat out for you anymore. And anyway why don't you ask my opinion!

Kind of reminds me of the story of the Chinese rail workers who 'must' show a set number of teeth when they smile or the US grocery shoppers who 'must' always be nice to their customers no matter what, hence causing a strike when workers felt they had to be nice to customers who might potentially be threatening!

My god! Not only are we in danger of taking the customer out of the loop but the employee – the best judge of emotional dispositions.

Hence, the firm with these tools does efficiency well and some elements of experience but also risks assuming that the data gained are the truth and nothing but the truth. This can only lead to a focus on spreadsheet mechanics and a sacrifice of employee knowledge, customer psychology and expertise.

And how daft is that!

I mean if Alex Polizzi notices the signage outside a restaurant is poor and can be improved she doesn't say, 'Oh! I shouldn't do that; there's no ROI on signage and that new colour scheme and brand logo looks good but I can't use it since we have no evidence of its impact; especially since none of our current customers are talking about brand logos.'

It seems to me, that if the Fixer was in fact an average business executive or management consultant then the show would go something like this. You have a terrible experience, cut your costs and put in this measurement system. Because what gets measured gets managed right! That feels to me like the surest way to end a business and the best way for management to avoid responsibility, that is, by hiding behind the number, an over-focus on command and control, an emphasis on 'It's your target, not my fault', and an increase in myopia: the inability to see as the customer sees (including those who don't use your brand) or 'could see'!

But it's more than that; it's an IT data-driven zeitgeist that doesn't know its own boundaries!

I mean, AI certainly offers some cool approaches, but it also makes false assumptions that humans are always calculating engines; that the subjective world is amenable to generalisation; and that everything in a system interacts in a fixed way, when in fact, how fishnchicken staff react to me affects my reaction back to them, and so on and so forth.

The DEM Model

So what are we to make of all this? Well, if we wrap this understanding into a simple model, we would find the following[3] (Fig. 2.1):

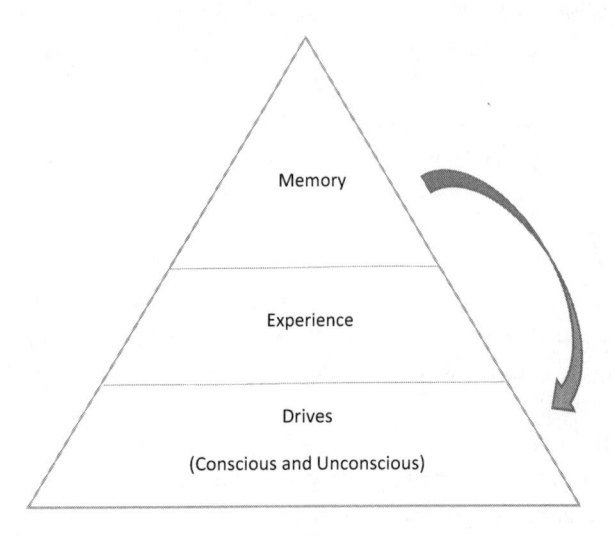

Fig. 2.1 DEM approach to understanding customer experience

Drives

As Dr Simon Moore explains; 'Drives exist at the deepest level since what *drives* the customer relates to their goals then their emotions and motivations – where their energy is directed. Drives can be intuitive and primary; conscious and non-conscious. They encompass needs and arise pre-planned and from the moment.'

So

'I need a coffee – my goal is to go get one (how) – I can't be bothered to walk too far today (motivation) so I won't go to my usual nice one (previous emotion) but the nearest and easiest (current emotion). ....'

'I had an awful experience in that corner shop when I went in last week to buy chocolate (emotion at decision moment) – I don't like that shop (emotion-behaviour postevent) but NOW I 'need' some chocolate and I can't be bothered to walk in this rain to the next shop...so hell I will just go in this one again....'

Experience Layer

Then once we engage with 'the experience' our drives are readjusted. For instance, we might find that staff in the store are surprisingly surly and rude or 'that the shop seems dull'; 'that mobile network is reliable'; 'they really care for you.'

Memory Layer

After this experience we then lay down new memories that readjust our drives, what Baumeister might refer to as the If–Then rules, that is, the way we might say: 'If I go there then I might feel good or bad, because of these experiences.'

Hence, it is these new memories based on how the experience affected us that we need to uncover; not an easy thing when in a survey we frequently post hoc rationalise and say 'It was down to the price!'

The Feedback Loop

And then there's how we remember the experience through time, how memory feeds our drives, is durable or nondurable and how firms affect memory through their ongoing interaction – or noninteraction – with the customer.

So my memory of MnS is coloured by how they cold-called me two years ago to sell insurance! And my memory of a higher interaction product such as my BT phone line is affected by how I couldn't get to talk to someone when I needed to, instead being pushed to Chat Live!, an interaction that didn't work.

But for me there is an even better way to appreciate customer experience.

'Real' CX

Customer experience is the smell of the LUSH soap as you walk past the door; the look of the Apple store designed like a hotel foyer; the clarity of a network connection when you have the right software on your handset; the empathy of the Virgin Atlantic staff; the design and theatre of the Geek Squad approach when they come to your door to mend your PC; the care of a Zappos call centre rep when they engage you on the phone; the personalisation of a DPD delivery when they show concern through their text messages; the interest and excitement of the Emirates 'Flight Simulator experience' in the Dubai Mall; the Harley-Davidson bike which is more than a bike but a social identity; the Hunstanton Sea Life Centre, with the entrance Aquarium; the flower in a VW Beetle, the Overbury onsite concierge-style experience for intermediaries and project managers.

It can be tactically designed, in the case of the Philippe Starck Juicer or it can be strategically designed, in the case of Starbucks and their Third Place experience – not just a place to drink coffee but an environment to sit down and talk to friends.

But when directly or indirectly connected to customer value creation, it is about something within the customer: a personal and memorable 'experience' that relates to what drives us and our goals at each moment. To quote the sanest of marketing gurus, Professor Byron Sharp, it is also about 'Memory Assets'.

So, customer experience is, in the moment, qualitative, phenomenological, subconscious, sensory, affectual, emotional and rational: the 'how' of things as received by the mind of the consumer, sometimes a personal and memorable moment that affects behaviour; other times less resonant but experientially important nonetheless.

Consider this, at the end of the day, if an experience does not affect *what drives us* it is no experience at all.

As Dr Nigel Moore points out:

> I think the point is that consumer behaviour (any behaviour) is influenced by current associations. These are the memory traces that influence current/future behaviour. System 1 memory traces are "recalled" automatically, system 2 are "thought about". If the consumer experiences an emotion at the time that is forgotten…i.e. does not form an association…then that emotion will not, indeed cannot have an influence on current/future behaviour.

What Customer Experience Is NOT!

It Is NOT About Experience Management Without the Customer

Most brands are not Experience brands, they are Efficiency brands. Their focus is not on creating a sensory impulse or personal and memorable moment; their concern is for zero defects, the management of the 'nonexperience', the prevention of loss aversion.

So, when the tap turns on, the water flows, when you seek a mobile signal, you get one, when you enter a store the floors are clean, and when you go to a website it works, in the sense that I don't notice any problems and it doesn't cause me to remember it negatively.

But for these brands and situations my message is, we cannot call something 'customer experience' unless it improves how customers 'think, feel and behave' towards your experience.

- So BigBelly trash cans in Philadelphia which save the city $900,000 per year by calling up the municipal authorities when they want to be emptied may be a great example of Opex reduction and process efficiency but it is not customer experience unless the dial moves in terms of how customers perceive 'the subjective experience'.

- A bank's cost-cutting activities that force customers to manage their bank accounts online cannot be called customer experience unless they move the dial in terms of how customers perceive 'the subjective experience'.

Each of these approaches is typically called 'inside–out'. In fact I would go further and just say no customer benefit – no customer experience.

And there is a real financial penalty to focusing on 'not the customer experience'. Huge sums of money wasted on hygiene, hygiene and more hygiene. Sure you can see data but they are valueless precisely because they have failed to consider the soft side. Or we spend oodles of money chasing slight reductions in churn when our efforts could be better spent on innovation!

You can't after all maximise relationship value based on things working as expected. Although, for sure, making things work can be 'an experience' if it is 'meaningful': as you might find in the Amazon Jungle if you get a signal on your handset! That is more than service delivery!

It Is NOT: A Tool

CEM tools are not CEM any more than Sales tools are Sales.

So don't confuse a tool's capacity to industrialise and scale with the process of doing the work!

Tools, AI (artificial intelligence) for sure deliver fantastic and realtime opportunities for creating experiences; they automate brilliantly but they can never replace the why, the hunch or what it means to understand the customer: except in a limited way.

For instance, I remember watching a video by one vendor on the benefits of realtime campaign management. Cue nice pictures of the customer walking past a store and getting exactly the right offer delivered at the right time to him. The experience effect was positive!

Now imagine you are targeted on sales growth. There is a risk you buy into this, and think every message put out, customers would 'love' and gains you revenue! But the simple truth is, you together with lots of other vendors are all firing messages off to the customer, and many of these are incorrect, mistimed, an aggravation and so forth. If you are not careful, if you don't consider how your messages are being received sure you will generate some sales growth from a few customers, but many more will end up disliking you and tuning out! These guys just send me spam!

We can see the same effect with the coming Internet of Things. In 20 years data will be constantly fed back on the performance of your clothes, your car, your health, your house. Can you imagine the blizzard of upsell messages that the customer will then receive on a daily if not hourly basis! Not to mention the security risks.

I predict that, in the same way as some digital natives are starting to junk commoditised downloads of books and music in favour of noncommoditised physical books and vinyl, the human touch will come to mean more. That doesn't discount the importance of IT; it just means that to cut through the blizzard of commoditised noise, the subjective response within the human–technology interface, must be considered!

It Is NOT: *The Result of Summing All Experiences*

At the recent Forrester CX Forum I met a Head of Customer Experience from a pharmaceutical company. Interested in her view I asked her what she thought customer experience meant. Her view was that everything matters, including how the cleaner responds to you. Now I can understand where this comes from. In manufacturing, quality is part of every stage of the process, but I would contend that when it comes to customers the inherent objective view that customer perception is the sum of experiences in the same way that a car is the sum of its parts, is not only wrong, but dangerously wrong.

In fact this is the biggest error in customer experience!

For customers simply don't sum all experiences! Customers are not cost-benefit calculators and do not construct reality based on noticing and weighting everything. They notice what is salient, and discount the rest. They are open to influences and change, in a way a closed product engineering system is not.

The challenge then is not to sum everything and control everything but decide what to focus on to engage the customer. To move the dial on the experience they have.

It Is NOT: *About Gaming Data*

Goodhart's rule applies here: 'Any measure that becomes a target ceases to be a measure'.

Typically firms have the objective of raising an attitudinal score to satisfy a bonus requirement or an internal political objective, which is therefore reached. Unfortunately what firms fail to do is ask 'how' it's been reached; which means a gamed approach to data analytics. Bonusing is therefore not a great way to drive customer experience. High leadership salaries and bonuses also threaten not just to game data but lead to low empathy and distance between you and your employees (the other wealth creators!). You should aim for CX to be an intrinsic part of your business.

It Is NOT: *About Just Moving Attitudinal Metrics*

When dealing with subjectivity any survey is only ever a proxy of reality. Let's face it, it has to be. The neurological connections in the brain and what they mean at the moment a decision (in the decision) is made are impossible to quantify and not fully understood even by the customer, who typically might say, 'Yes, it was OK', or post hoc rationalise a response: of course it was the price!

This means that all research can only ever be informed opinion. And attitude or emotional metrics are fleeting, changeable and easily gamed.

However, that doesn't mean it has to lack technique. What it does mean is standards have to be in place to avoid bad practice. And bad practice can be seen with measures that depend on response bias to deliver a result, use scales that miss out and underestimate the many perceptual and other types of experiences that motivate behaviour, and apply statistics that assume subjective data behave in the same way as machine data.

It Is NOT: *About Physical Things Only*

We are talking drives here not the customer went 'on the website' or 'entered the store' or even 'received a bill'. That's not a drive; that's a physical experience ('an activity', 'occurrence', 'interaction', an 'efficiency event'). Anything tangible must therefore come with intangible benefits.

So mapping physical events alone is not mapping what drives the customer. And a company that does the former risks ending up with a process map where every piece of technology is 'important for the experience the customer has'; the only thing is, there is no customer!

Personally, I think total quality management (TQM) is at fault here. TQM works fine when dealing with machines or service processes, but it should be obvious that applying TQM into designing the 'experience the customer has' is wrong-headed.

It Is NOT: *Customer Service*

Customer experience looks from the point of view of the customer, seeing as they see. But it goes beyond notions of service in the way we seek a deeper understanding of what drives and motivates customers: how we can build 'memory assets' and how creating an experience could attract custom. Hence Stew Leonard's sell grocery products in an environment that expresses beautiful design and 'dancing bears'! Things not measured by SERVQUAL or considered by many service designers.

A simple way to appreciate this is through the CX equation:

ExQ (experience quality) = f (PQ (price quality), SPQ (service, product quality), CJQ (customer journey quality), where f is based on what drives us based on our goals and subgoals. This is something we go into later.

It Is NOT: *A Panacea for Every Situation*

Experience brand status is a competitive differentiator not a rule of thumb. There are plenty of brands in my bathroom cabinet for instance that I would call Efficiency brands: Bausch and Lomb, Gillette, Mum, the list goes on. They have positioning, a habitual buying status and probably don't need experience. That is fine. However, for a new entrant, a brand extension, or where there is commodity pressure competing on the experience is an option.

What Is Customer Experience Management?

Customer experience management (CEM) is the management of 'the experience the customer has'.

Hence, we cannot really talk about management in the traditional sense of the word. The customer's personal world is not amenable to an engineered approach in the same way as say fixing a car. So although a machine would deliver quantitative and objective information in a fixed manner perfect for a command and control style of management, by contrast, subjective data would be less about root cause and more about how things are qualitatively perceived, how things emerge, change and interact.

Hence, to apply a management rationale suited to control would mean we are mostly doing efficiency: hunting down instances where objectivity is apparent in subjective data; focusing on the 5% of value from reducing defects and missing out on the 95% of value creation through creative design.

For some people this has meant we should throw our hands up and say, customer experience is different, more qualitative and personal therefore we will ignore it or let's focus on something we can control. Whereas for others, this has meant we should only focus on 'getting things right', experience as in what objectively happens.

I think this is wrong. We should not be afraid to seek management principles when faced with the challenge of handling customer subjectivity. And for me, although customer experience may not be so amenable to a control environment, it is possible to influence *the experience the customer has* if we consider the term 'management' to also mean design, setting parameters and agility.

Hence, in this way, customer experience management becomes as much an art as a science.

So although it is in part consistency and control it is also about how we can guide customers; respond to their changing drives; ideate, design and test experiences; engage in cross-silo collaboration; use the voice of the customer data and embed in our organisation the ability to understand what it is like to be a customer. In CEM the principle for me is that we 'go with the flow' and proactively respond to opportunities as they arise: delivering more personal, memorable and targeted experiences.

Here are some examples:

Julie Walker is a leading social media expert and guru. Her firm Purple Spinnaker has advised some of the world's leading companies on their strategy. Recently she overheard one conversation in a golfing club. Here female golfers were complaining about the heavy and badly designed umbrellas. How the female golfer was not catered for. This led directly to the development of Golfing Gertie, a social community and outlet for golfing umbrellas and products for female golfers. In effect the firm was set up on a hunch in an agile manner, based on a close understanding of what it is like to be a customer, not an analysis of reams of data.

Another social media example is how in my own work, rolling out global best practice social media platforms, I used this 'agile' data layer to pick up the unprimed commentaries of customers on network, marketing and customer care issues. This was then used to target potential key influencers on network perception and identify where customers were talking about the network away from tweets hash-tagged to customer care. In addition, we built the capability to assess 'how customers were talking about their engagement with the operator' that could help in the design of new marketing campaigns and push notifications through customer care.

A final example comes from the hotel trade. Here faced with a commoditised market, hotels are seeking uniqueness by designing in experiences that are personal, memorable and unique. One hotel has a quirky approach, positioning Red Penguins in and around the rooms; another offers hub rooms, small pod-like hotel beds that serve up the ideal experience for that subset of consumers who don't want a fancy hotel. Here management is less about summing up touchpoints and preventing points of failure and more about creative design, trial and test and seeing what works but nonetheless putting customer experience at the centre.

Hence to manage we need an innovative space and cross-functional teams that can execute change. We need to listen, get closer, to the employees, suppliers and customers who live the experience, close the loop with realtime responsiveness that builds relationships not just fires off a sales message and perhaps above all else set relevant metrics and understand 'what we want the experience to be'; to quote Shaun Smith, to be aware of our purpose and our brand theme.

If you don't manage customer experience like this and end up discounting the human parameter, then you are not going to create an Experience brand.

And let's face it; there are plenty of businesses that do just that, while still proclaiming they 'do CX'.

For instance, consider the CTO who prevented the redesign of a retail website based on customer need due to 'the lack of demonstrable ROI'; or the sales executive who put a timer on purchasing services on a website because it forced closure of a sale. Or the regular way, in which social communities are sold: not to create a better conversational platform or the ability to create more personal and segmented experiences, but the opportunity to deflect call centre costs; hence reducing engagement.

In each case there is no doubt a valid sales-generating, cost-cutting reason for these actions. But this is not CEM because CEM is not only about finding revenue-generating opportunities but also about eliminating activities that generate sales or cut costs if they are to the detriment of the customer: which means the governance process has to be about being the customer's advocate.

After all, in an environment where:

The CMO is out to get as much money from customers as possible
The Sales Director is out to sell as much to customers as possible
The CFO is out to save as much money in dealing with customers as possible
The CTO is out to spend as much money on IT equipment as possible
The CEO is out to maximise shareholder value based on sales
The Insights Director has no voice except to support sales
The Service Director is out to minimise complaints
The HR Director is not engaged except at a functional level Where's the employee experience!

Someone has to represent the customer's interests, 'at board and strategy level'! Someone has to have the cross-silo view. After all customers don't view companies as separate silos when they engage with you; they don't care; they just see the experience!

Which of course is a huge challenge to the firm because governance is much more likely to be siloed around products not customers; with no PnL account for the customer journey; limited ideation and poor cross-silo functioning.

And such a governance approach must be set at strategy level to show intent. Which does not mean it has to involve the creation of some big structure where a CE Council over-rules departmental activities!

Indeed, much of the activity of governance can be about being an evangelist between departments, sharing and interpreting data and creating a virtual design team with a remit and budget to show improvements in the experience. Such as how RBS executes customer experience, focusing a team of 20 on troubleshooting 'experience improvements'. Or how exemplar Experience brands focus on culture, regularly visiting and engaging parts of the organisation in 'customer experience'. And calling out examples where activities are diametrically opposed to a customer experience mindset.

Or in a travel company, how innovation decisions were fed through a key cross-functional team that included IT. This team made the call on whether to invest, hence achieving buy-in from the start.

Governance also needs a strong thread of challenge from outside the industry. I for one believe recruitment of CX professionals should not be bound only by industry knowledge but by CX knowledge.

Here then is what Michael Young, leading CTO and CX expert, says on governance and why it's so important:

> The path of least resistance is always picking off the single issue or challenge in a company because it's easy to identify and plays into the politics or structure of a company ... have a less than stellar performance in retail; let's focus on the stores .... Have sales dropping; let's focus on marketing ... Have products getting returned, let's focus on manufacturing etc etc etc ... and as we both know none of those things focused on may resolve whatever the problem actually is. That's why I think UX becomes an easier play. Companies need the thought leadership at the top required generally to drive CX ... one of the reasons may well be that there is a lack of non-industry execs at the top and also because they've had captive markets for so long.

Who Is the Customer?

Finally in our right understanding we need to answer the question, who is the customer and who are we! I mean, if we are interested in the CTO's customer experience then maybe we should major on Opex reduction. By contrast a CX Director may be interested in how we can help the end customer, in which case we need to understand the end customer's viewpoint, what drives her. And this is not as easy as it sounds! Executives typically fall into the following traps, each of which takes them away from understanding who the customer is because they use themselves and their job roles are the benchmark!

The Myopia Trap

This is where the customer is not 'the customer' but in fact 'your company or department'. So customers are only concerned with price (they think like the finance or marketing department); they are only concerned with in-store service (they think like customer service); they are only concerned about download speed (they think like engineers). And each one will be supported by insight! This is something I have at times challenged, for instance, using the language of engineers to ask them to consider the Network Green–Customer Red problem or critique research processes on poor and myopic methodology that misses the bigger picture by just serving up what the internal stakeholders want to hear.

The 'What Looks Good in the Executive Meeting Trap'

This is where the customer is not 'the customer' but in fact 'you'. Here you will not be concerned with what the customer thinks or feels but with whatever approach gets you through the executive meeting. We all know what this means: get a big name vendor to provide the system and data and 'game' it for you. We play the NPS game, as one conference delegate put it.

The Tangible Trap

This is where the customer is not 'the customer' but in fact an IT system. Here customers are assumed to look at physical touchpoints only. Intangible drivers play no part in decision making at all!

And Who Are We?

And then there is the important question of who are we?

In exemplar customer experience companies the brand purpose (the theme) is linked to the customer experience and the experience is the measure of the brand. So, the Apple brand is linked to innovation, Metro Bank to trust. In each case, the way the experience is designed reflects these qualities. And these qualities all imply a focus on soft metrics to support your Experience brand concept. For instance, go into any Metro Bank and the design has a more personalised, warm and trusting feel. Pens are not on chains; school trips are provided to visit the vaults.

Be clear though, branding the experience is not a trite question. Just saying, 'We want to be the most trusted brand' is useless if this is not reflected in how you 'motivate' consumers around the trust theme: and this equally applies to your employees as well.

Likewise, if you answer the question, 'What does customer experience mean to your brand?' by saying it means 'Summing all touchpoints so they are above expectation' then that's what you'll get! And for me that is a very inside–out way of speaking because it reads: a tied in knots, IT heavy, systemised and process-orientated experience concentrated on reducing loss aversion but doing little to motivate consumers especially when consumers don't expect everything you define to be above expectation.

And anyway in your answer you haven't told me how your brand feels different to me as a customer; if you want the customer to say 'above expectations', how and why? What's the new customer experience? If I was at a party, what would I say about your brand that's new, differentiating and drives me?

Right understanding is therefore critical; all else flows from it. If we start from the basis of a misunderstanding then our objectives, measures and implementations will fail to deliver customer experience improvements. So far from being fluffy, understanding what customer experience means is the most practical starting point of all.

Management Implications

1. CX is bastardised; make sure when you use the term customer experience you know what you mean.
2. Customer experience means: 'Those subjective experiences that lie in memory or are experienced in the moment, that influence and are influenced by our drives which in turn lead to behaviour.'

3. Customer *drives* underlie the experience the customer has: understand them and design for them; don't just think of creating memories that may lack durability.
4. Drives encompasses expressed and unexpressed needs as well as things the customer may not have considered. Mapping goal and subgoal states is critical to appreciating drives.
5. The real benefit of CX is it makes us think of the phenomenology of the customer; it is not just about emotions or service delivery.
6. Many measurement regimes have been designed by engineers; understand what you are missing.
7. The DEM Model outlines the key concepts in experience.
8. Be aware of what customer experience is not.
9. Managing '*the experience the customer has*' means managing for subjectivity: this is less about control and more about design.
10. CEM is about being the customer's advocate when other stakeholders are focused on firm benefits only.
11. Identify the customer and be aware of your purpose/theme.
12. Understand what we would like the customer to say about us in comparison to competitor brands.

Notes

1. Hence the old adage that claimed drives only weakly correlates to spend behaviour is true, but misses the point.
2. Taken from Dave Snowden of Cognitive Edge.
3. I avoid the transformational experience which goes beyond customer experience. But point the reader to the book *The Experience Economy*, by Pine and Gilmour, to understand what this means.

3

Right Commercial Principles

'What is the return on customer experience?' is like the question 'What is the return on innovation?' To which the answer is, it depends on what you do and how you do it! The important point is, nobody says, 'Let's see if there is an ROI on a prototypical word like innovation before we innovate!' And the same should apply to customer experience: that is if you are serious about it.

After all, customer experience isn't a choice, there's always an experience whether you want to manage it or not and the ROI of customer experience is simply about the return derived from investing in the customer as a source of value just as we would a product or service.

The ROI of customer experience is also the wrong question! CX after all is an enabler; it gives the firm strategic options. Hence the question is like asking: 'What is the ROI of having a phone!' (Source: Professor Stan Maklan, see webinar on www.allaboutexperience.co.uk)

Which leads us to a far better question: not 'What is the ROI on customer experience' but 'What is the ROI derived from the different ways of doing customer experience? And which way is best?'

However, before we get into that, to do customer experience commercially we must appreciate a core principle.

The Core Principle

The economics of customer experience are about how we can add value to the customer's account with us, not how we can add value to our business account with the customer! The fundamental underlying assumption being that by adding value to the customer they will add value to us, through extra sales, more use, more spend and longer tenure (Table 3.1).

So, although direct mail may well lead to higher sales volumes, if it doesn't improve the experience the customer has, it just becomes a better way of sending junk mail. Something that I would argue does not build an Experience brand.

Likewise if we just focus on selling the cheapest priced phones while at the same time creating a store environment that reflects a focus on discount merchandising and avaricious sales teams we may find that the supposed price benefits are impacted upon by the experience: how customers think and feel.

Hence, if customer experience is of strategic importance, companies must focus on improving the experience the customer has as part of their DNA.

But in all honesty this isn't such an easy thing to do.

Why? Because customer experience being subjective, involves engaging in 'actions' that many times involve moving small things that in combination form an experience, something Michael Young, CTO and customer experience expert, calls 'the cocktail'. Hence, you can't always rush to justify your experience actions through some ROI, and if you do your modus operandi risks being only about closing off errors and defects that have a known and historical aggregate impact on performance or you will aim to drive up sales without consideration to the customer view.

And in this way you will build an Efficiency brand not an Experience brand. You will use the techniques of service delivery – which is nothing new!

Hence, customer experience means investing in the customer, adding value that hasn't been there before and is sometimes difficult to quantify without a trial.

Table 3.1 Customer account and firm account

Customer account	Firm account
How the customer thinks and feels	How the firm thinks and feels

3 Right Commercial Principles

Which of course leads us to a frequent paradox of customer experience … We only ever see the benefit of our CX strategy after we do it; and if we want to predict the benefit beforehand we may end up engaging predominantly in efficiency.

So if you're the type of company that would prevent investment in blankets for patients because it costs money and you can't see the importance of creating a bigger sense of customer care through the many actions we perform then I suggest you have no chance of ever creating customer experiences. Likewise, if you're the type of company that pushes a sales message of a bigger boot to your car, while at the same time cutting costs by removing the spare wheel – hence causing a major flashpoint if my tyre bursts – then you are not an experience company.

Of course Experience brands do not have this problem. Investment decisions are based on creating customer value. And they understand the need to trial and test: 'Screw it, let's do it!' as Richard Branson famously said. For instance how did Apple know that the look and feel of the Apple store would help generate value? The answer is, they didn't; it was part of their Experience brand value!

They understood the source of differential value was the customer!

They also understood that if CX investments were done correctly, any competitors would have to follow suit or risk losing market share.

Hence, traditional firms focus on efficiency benefits (FX) while CX (CEM) firms focus on customer-based differentiation: up to a point FX delivers benefit but after a tipping point it no longer works as a competitive differentiator and CX takes over! Clearly, if you're not ready for that beforehand, you're sunk.

So the FX firm will say, 'We'll get to CX next year!' confuse CX with efficiency and follow a vendor and consultancy strategy to sell software and IT consulting. After all it's easier to advise on efficiency than to innovate around the customer. Empathy is sorely lacking amongst these stakeholder groups.

However, the CX firm will activate parallel strategies and understand the difference between the two.

Hence the best strategy is to work on both. We must not kick CX into the long grass.

I might also add that another benefit from this CX–FX approach is that we gain a clearer understanding of any hidden relationship losses resulting from short-term cost cutting and sales growth approaches! In other words because we know what CX is, we can use it to test out if such strategies are causing losses in the experience the customer has whatever the immediate firm benefits.

But be warned, all this takes effort!

For instance, most companies will want to know immediately how they can monetise the customer experience. But if CEM at its heart is about differentiation through developing a deeper relationship then we have a problem because relationship takes time, relationship is about give and take, relationship is about investment, and relationship is about listening to the customer and helping them, not rushing to the short-term sale. Relationship involves not sacrificing sales today for sales tomorrow; relationship is an eco-system involving your employees and your suppliers. Relationship value is not about the short term. Relationship is about understanding what sort of relationship we want! Realising that many times customers don't want to be your fan, they just want you to provide solutions!

Are you ready for that?

I am not so sure. It seems to me that most companies are obsessed with Opex reduction and couch CEM in those terms. Hence, we find companies saying:

We can improve the experience the customer has by cutting costs! For me, that as a primary CX objective, is plain wrong.

Or they are obsessed with short-term sales growth: we can improve the experience the customer has by selling more to them. That's just a sales plan.

Or senior managers become too abstract in their understanding, believing improvements in the experience the customer has means NPS increases. Oh really? Just because a number changes does that mean the experience has become: more personal and memorable, more meaningfully different, driven loyalty and relationship, led to more use more spend in a positive way? Or, is even durable!

Perhaps I give you a higher score because the experience is more hygienic and efficient but means nothing more than that, especially when say a +10 NPS score increase hardly means any statistically relevant improvement in average recommendation.

Stop chasing the metric!

Now I know what you are going to say. But efficiency is giving the customer benefit. But for me, although efficiency is good, it is at best only doing business as usual. It is not a new paradigm. Even worse, it tends to get significantly conflated with not only service but another term, UX (user experience).

Now I know UX is supposed to be about designing products and services with the customer in mind. But in reality UX often remains myopic in its focus on efficiency and fails to consider the broader customer experience.

For instance, network perception is influenced by brand and billing but design remains doggedly siloed.

Hence, I have a problem with it! UX to me seems to being defined as 'Use' experience rather than user! Make it work! Not create an experience.

The Return on Customer Experience

With that in mind then, let's see what types of experience deliver the highest return. These can be divided as follows.

Lowest on service efficiency (You should be doing that anyway.)
Higher on service excellence (I'll pay more for that.)
Higher on finding new drivers that innovate the current goods and services (I'll redistribute my share of wallet for that.)
Highest on finding new drivers that innovate the experience beyond the current goods and services on offer (I'll pay more for the third place in Starbucks or the aesthetic experience of soap at LUSH than I would for the goods or services alone.)

I call the management of these layers – efficiency (or assurance), excellence and drive (or sometimes I say motivational)[1] – subjective asset management (SAM), the benefits to the customer of which are summarised in Fig. 3.1. I go into more detail on these layers in the following chapters but to illustrate the point of SAM let's take an example from the telecommunications industry.

Efficiency: Means we focus on enabling a connection. This is the layer where we manage the 'nonexperience'.

Excellence: Means we focus on enabling good service at all times. Making sure marketing doesn't rip us off, we have a noticeably good signal, a noticeably seamless UX and customer care is responsive and empathetic. This is a layer where we create customer experience.

Drive: Here the operator provides experiences that affect customer drives (both conscious and unconscious). So, there are Experience Hubs in stores where we can 'experience' 5G network performance; handset design is aesthetically appealing; we receive some great experiential and personalised offers through text messaging, OTT services such as apps or we receive care moments in the journey such as warnings on high data use. Here, customer experience becomes its own economic offer, part and parcel of something beyond service delivery such as delivering a sense of 'we care'. And also something that is noticeably different from the competition.

In the above example, we can see that we only add value to the customer's account if we add value to their subjective, personal and memorable experience: apparent at the excellence and experience layers. Not denying, of course, that we can lose customers if efficiency (a service quality paradigm) is not maintained!

And we move between these layers due to commoditisation.

So efficiency is just the ticket to the game; we are preventing loss aversion but we are not moving the subjective dial.

Once commoditised we move to excellence to achieve higher returns on the experience the customer has.

Finally the commoditisation of excellence leads the firm to build experiences (see Footnote 1) which deliver the highest return because they target the consumer's deeper conscious and nonconscious drives (needs).

Ultimately though even these experiences will become commoditised as new innovations at the product and service layer scale and we are taken back to a new parallel round of efficiency.

Companies that succeed then are those that can manage this cycling over time by embedding a customer experience approach into the way they manage and measure the experience: understanding that experience is not just about efficiency, unless they want to be the lowest-priced provider.

3 Right Commercial Principles

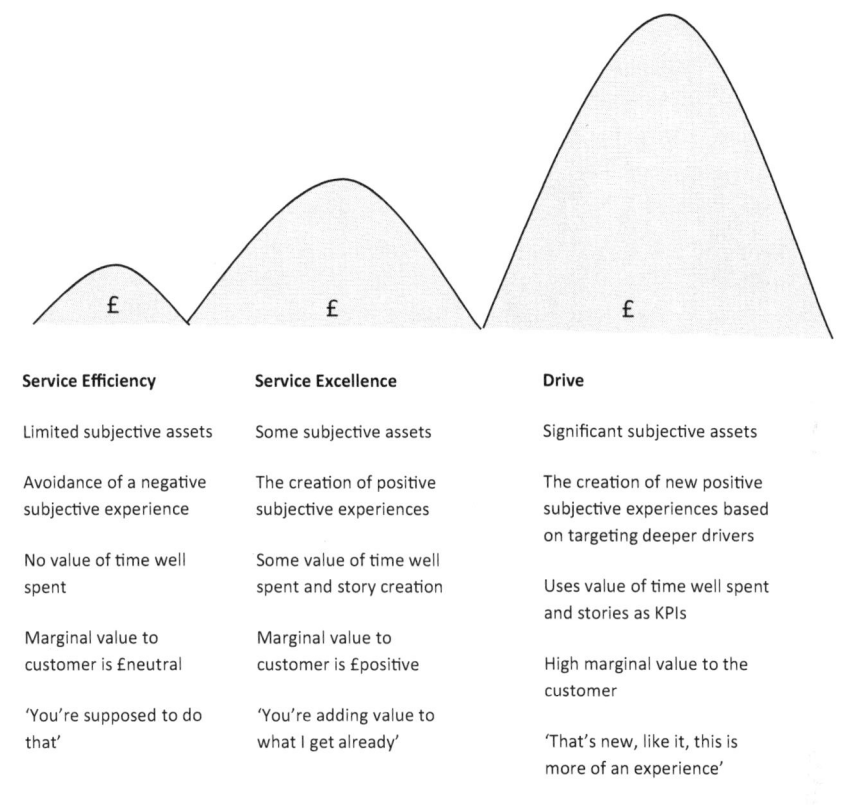

Service Efficiency	Service Excellence	Drive
Limited subjective assets	Some subjective assets	Significant subjective assets
Avoidance of a negative subjective experience	The creation of positive subjective experiences	The creation of new positive subjective experiences based on targeting deeper drivers
No value of time well spent	Some value of time well spent and story creation	Uses value of time well spent and stories as KPIs
Marginal value to customer is £neutral	Marginal value to customer is £positive	High marginal value to the customer
'You're supposed to do that'	'You're adding value to what I get already'	'That's new, like it, this is more of an experience'

Fig. 3.1 Subjective asset management in a service

We also see from this how commoditisation itself is a route to innovation!

We feel pressure to differentiate which leads to the creative impulse!

Interestingly though, this is not a linear effect! We do not go from product–service–experience. We move between these layers as sources of differential value, relevant in some cases, irrelevant in others. So, in my opinion, going back to that classic example of experience value, 'the birthday cake'; commodity cake providers make the cake-making process easier and more fun. We see TV programmes such as 'bake off' and some of us go back to baking a cake and making our own party, just as others stick with the purchase of that McDonald's party with birthday cake provided.

For me innovation is not only about 'experiences', but also about product and services. All exist in the one environment! There is no contradiction, only context.

The Challenge to Traditional Business Processes

However, we should not underestimate the problems of executing SAM. After all, experience is a huge challenge to create and manage because it depends on the idea and the execution and requires a close understanding of how customers think and feel.

And the simple truth is, our traditional business processes are not up to the job!

For instance, insights might say the price of the airline flight, the flight times and seat comfort are the most important driving factors in purchasing a ticket. But such surveys are reflective of post hoc rationalisation and an abstract response outside the decision moment. In short, they fail to capture important effects in the ongoing experience such as how on one flight, the broken entertainment system is a problem; the next flight it's not a problem but the inability to put my luggage in the overhead lockers is.

Hence, in reality, although such things as flight times and seat comfort are critical, they may be good enough already or difficult to innovate, which means what really is 'meaningfully different' for experience creation are those aspects of customer care in this case under-reported or not reported at all.

And anyway from our first example, what is a unit of comfort and if you do make them more comfortable, how does this affect the rest of the 'subjective' experience the customer has?

So, in context, the great quality food in economy class in Qantas is memorable and valuably different for me; but this also means I am more sensitive to those 'experience quality' differentiations that work less well, as well as, of course, the human aspects of the experience which do.

I would argue then, that unless we fix our 'broken' understanding of customers we cannot hope to move on and design for experience.

Which brings me back to Experience brands; how do they 'understand' and execute experience?

Examples of Experience Brands

Well it seems to me that they have a better handle on what it is to live the experience: they are not in awe of a gamed statistical model, have a better understanding of who they want to be and why, and are more willing to take the risk. They have creative equity; they design *for* experience, so:

- Cerritos Library doesn't just lend books 'excellently'; they use the library as a platform for an educational experience: hence engaging with the drives of their customers to be entertained and informed. They sell entertainment before they sell a library.
- Starbucks engages with customers at a deeper level by selling more than coffee excellently, focusing here on using the occasion of coffee drinking as the platform for their 'third place' experience. They sell the third place before they sell coffee.
- LUSH creates a deeper sense of engagement by building a business around the aesthetic appeal (the experience) of soap and how it can be used. They sell a sensory experience before they sell soap.
- Zappos and First Direct take the call centre experience and focus on how it can be used to engage customers; they make it an experience when you phone not just something that is delivered well. They sell care before they sell an account.

Of course, this doesn't mean we do anything and everything for the customer. There has to be a return. For instance, one car recovery service in the United Kingdom undertook customer experience taking any driver caught in trouble back home for free at any time! Nice idea, hardly commercial.

And once these deeper drives have been targeted then the brand designs an experience around it. The sense of theatre created then emphasises and 'sells' these deeper drives; this is the immersion part where we start to sell the 'experience' itself.

Not forgetting of course, that all the while the Experience brand company has its finger on the pulse, using customer data (which is not a trite thing to capture) innovatively in experience design while at the same time leading with experience designs.

Because this SAM approach is so fundamental to customer experience economics, let's take some time to go through in depth each of these layers.

Customer Differentiator: 'Get It Right for Me'

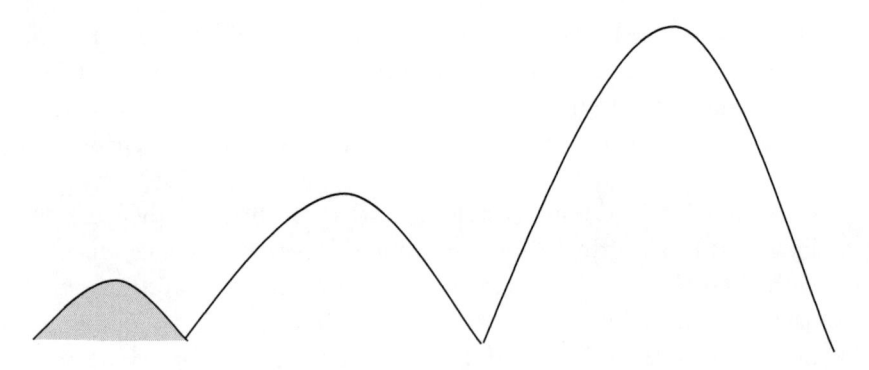

Fig. 3.2   SAM Tier 1. Service and product efficiency

Here SAM is about preventing loss aversion and ensuring that the features of the product or service work. Customers would say, 'Yep it's OK'. However, customers would not pay much of a premium for you to be in business unless the market is so poor that efficiency counts as service excellence. Hence, although this is a critical piece of the customer experience jigsaw it also delivers the lowest return to the business in terms of creating subjective drives and memories.

You do not get to be an Experience brand by just focusing on efficiency and preventing loss aversion.

This is where most businesses apply the term customer experience; assuring for instance complex multichannel and digital environments, removing pain points and reducing complaints. Talking about experience in a hygienic way: after all if your customers can't get into a shop because the door won't open, or it's too far away, you're not going to get much business.

However, if we remember that our asset is about targeting the deeper subjective drives of customers and creating those sensory experiences, it is difficult to see how there will be much in the way of an increased return on customer experience here. 'I for one would not pay an extra margin because you are the best at managing things when they go wrong!'

Efficiency is a requirement not a value add.

However, no efficiency means no experience so it is important to make sure the goods and services you provide are designed with the customer in mind. Indeed, by way of warning, here are some simple examples of service and product efficiency that did NOT do this:

- The large pharmaceutical company that bought a 'single view of the customer' solution. This turned out to be clunky and onerous for employees to use: the intangible employee experience had not been factored in so it failed.
- The telecommunications operator that cut their help lines forcing customers to use Chat Live!, immediately isolating any customer who had a complex problem or wanted to speak to someone, a particular problem for digital immigrant groups such as myself.
- The aggravating problem of trying to pay for parking over the phone when the machine fails continuously to recognise your voice: parking for 6 hours becoming parking for 60! And no customer service in sight.
- For employees the regular problem of 5 hours of your working week spent filling out your timesheet and other electronic forms, time that clearly was never accounted for in the cost-saving benefits.

So, to be efficient make sure you get the voice of the customer (employees, suppliers and experts!) into design.

And this data need to be monitored. For instance don't just depend on customer complaints which miss out on the many complaints customers don't report (silent attrition)! Complaints, however small, should be celebrated. Innovations should be encouraged.

It is fair to say that efficiency is also something that can work with machine data.

- Many telco network providers have operations centres that pick up any signs of cell degradation through reduced speed of web browsing. Because these are picked up early and relate to VIP customers' locational data the company can quickly resolve the issue before it spreads to

neighbouring cells. Loss aversion is averted and more spend more use is enabled.
- Analytics that focus on what people do, how they move through a website, are critical for understanding if problems are occurring.

Likewise, experts can help with efficiency by immersing themselves in the experience.

- This is how in my work with a travel company we picked up the importance of the port experience even though the firm was devoted to assessing the quality of the ferry journey itself. One outcome of this was a programme to improve signage. Hardly a point for generating memorable and sensory experiences, but certainly important for loss aversion.

So far so good then: Think Customer! That's a clear objective. But there are a number of risks with efficiency that we should be aware of.

Risk 1: Don't Rush to ROI

Going back to our ferry experience, you would have thought that these insights would initiate a series of investments! But this is not necessarily so. For, once the feedback had been received, senior executives made the mistake of trying to answer the question, what is the ROI of cleaning the toilets and good signage to the port? They had rushed to ROI: a real risk with Efficiency brands!

Of course because these are indirect effects difficult to quantify and part and parcel of a good place to do business in, it was not possible to answer in a root-cause way and the programmes were killed.

In other words, the rush to ROI missed the point that without an assured port experience the onboard ship experience was reduced in impact. Dissatisfaction with the port also acts as a lagged indicator to reduced value over the longer term and quite frankly a missed opportunity! Clean port experience means a better platform for nice restaurants and higher value of time well spent.

For me it's also interesting how no one ever asks the question, 'What's the ROI of unclean toilets and bad signage!' After all there is 'always an experience'. Or if you have a boring experience, 'You'll have boring drivers to that experience.'

The question is also 'a tell'. What it really means is, we haven't bought into customer experience as a strategic direction so we are really not interested in executing it and we don't care about building a relationship with our customers or employees! Hence, the programmes were considered a cost not an investment in 'customer experience differentiation'.

The question, 'What's the ROI of customer experience?' was not answered by the reply: 'What's the cost of not doing it!'

Risk 2: Experience as Everything

The next risk is how, if you don't include the customer's perception, you could end up calling anything customer experience.

This approach follows what I call the *little e* definition of customer experience. An approach best summed up, probably unwittingly, by Bernd Schmidt who wrote a full six years after Pine and Gilmore. Here customer experience is:

> The sum of all experiences a customer has with a supplier of goods and/or services, over the duration of their relationship with that supplier. This can include awareness, discovery, attraction, interaction, purchase, use, cultivation and advocacy. (Bernd Schmidt, 2003)

And what's wrong with that?

Well, firstly, customers don't sum up all their experiences to make a value judgement. The market research profession and psychology from fMRI scans to simple surveys would not agree with that. To state the obvious, we are not cost-benefit calculators who give a computer scoring on quality at every touchpoint (whatever that might be) then add these all up to come up with an overall rating! We use heuristics to make judgements, we pattern match, we use first-fit not best-fit approaches, we use emotion and reason, we have a subconscious and we are heavily influenced by others' opinions and authority bias. What is encoded in long-term and short-term memory, it seems, is not defined by an algorithmic calculation but depends on personal 'meaning' and context.

Hence, NOT everything matters.

I mean, if a tiger walked in the room, you wouldn't weigh up the pros and cons of what you should do: you wouldn't think; you'd run!

When I buy groceries, I don't say judge the experience based on scoring every part of it: let's score the entrance experience 5 out of 10, the checkout 7 out of 10 and so on and so forth. Never mind the fact that these have different contexts; it might not matter if the entrance is just a 5 out of 10 experience! And what are you going to do with the entrance anyway? Give me a bunch of flowers and make it 8 out of 10? So what, I won't spend any more with you.

I call this the 'can't see the woods for the trees' problem of customer experience. Breaking experiences down into ever granular metrics may force a response from the customer (response bias), but that score fails to account for the fact that humans don't construct reality in that way. They construct the world based on what drives them. And I don't mean they add up these either!

Granular metrics make efficiency the only obsession. As does the repurposing of the term 'experience' to mean 'everything objectively experienced!' However, I do accept that focusing on what people do objectively, as in seeing how they move through a website, is useful for CX, as long as we know that there is a relationship to the experience they have and don't assume it: that is, we know why and marry hard and soft data.

Nonetheless there are some companies that have been successful at mapping and considering drives in their efficiency design rather than just applying a sum of everything approach.

- One US airline designed their flight turnaround times to the experience the customer has, using social media analytics. Hence on one occasion, noticing that business executives in New York tweeted more often about delays less than the standard threshold of 1 hour, the company decided to invest in these sensitive routes so they would have a 30-minute turnaround time instead.

Risk 3: Fixating on Fix

Another risk with efficiency is that it leads to a fixation on fix. So sure, although we must fix the basics to achieve the least waste of time and effort,[2] we must also not stop there, which is unfortunately what frequently happens: companies saying, 'Oh! we'll get to customer experience in a year or so; let's look at our complaints data first.' For me, that means you have no skin in the customer experience game.

So be aware of this myopia in your CEM programmes. Don't assume that actions to improve on what is traditionally known as delight or transform cannot happen alongside fix. Sure they may only be small in scale while you fix the basics but that does not mean they are valueless or mutually exclusive (see Footnote 2).

In addition, alongside fixing things you can also look for areas of under-performance that customers have sacrificed to and may not complain about! Things that you can spot by comparing to exemplar experiences from other firms frequently outside your industry; this may be as simple as how the website looks or how its content is expressed.

Indeed, happy experiential moments help to moderate the negative effect of any bigger fix issues that might take time to resolve. And these kinds of experiences don't have to cost the earth.

For instance, in my work with one company, I found that 50% of the problems and opportunities with the experience came from simple communication changes and we could create wow and engage without too much effort or cost.

Another example of this approach can also be seen in the Nudge approach. How the UK Tax Office achieved a 10% increase in tax returns simply by changing the envelopes from white to brown to give them the appearance of greater personalisation. Or how one European company reduced tax avoidance by putting an 'eyes are watching' sign by traffic lights, using the association with rules and the colour red to good effect.

Hence, one of my challenges whenever I engage in experience design is to ask behavioural psychology teams such as Innovation Bubble, as well as customer groups and employees to walk the experience and come up with some simple and cheap ideas that could maximise the experience factor.

So to end…

Efficiency does raise the stakes. The ever closer attention to least waste of time and effort forces other companies to follow suit. Hence, it is important for commoditisation and moves us to look to differentiate and innovate.

Efficiency is part of the 'experience the customer has' because it avoids loss aversion (through least waste of time and effort) and sets the platform for CX.

But to define customer experience itself as this would confuse the firm into thinking they are doing customer experience when they are not, simply because they are not looking at deeper drives, creating experiences or using customer perception data effectively. The customer clearly has experiences beyond hygiene!

Risk 4: Efficiency for the Firm Not For You

I have alluded to this a little in the text. Least waste of time and effort can mean being efficient around the customer (otherwise they can't do business with you!) but can mean being efficient from the firms point of view – no customer included! Hence, the call centre operators who hold targets for number of calls responded to – cutting you off mid sentence to boost their perceived efficiency! Or the firm that moves everything onto Chat but fails to answer the customers query. It looks in aggregate that cost savings have been made to deliver the same service, but this is a false economy when inefficiencies remain for the customer.

Efficiency Is the Platform: CX Is the Relationship

Customer Differentiator: 'Get It Excellent for Me'

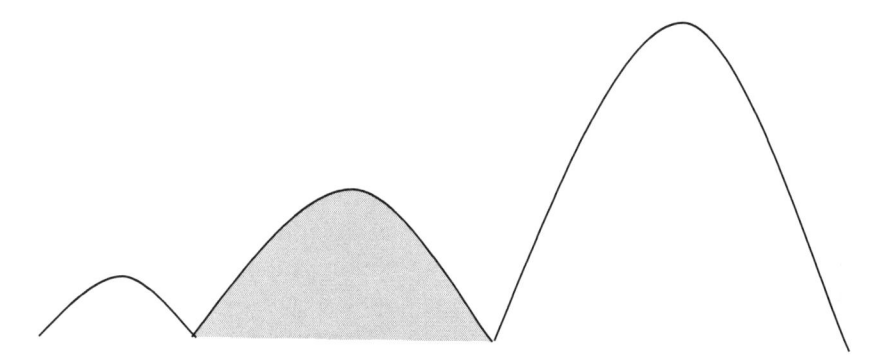

Fig. 3.3 SAM Tier 2. Service and product excellence

SAM is about adding positive value to the product or service sold; customers would say, 'They offer an excellent product or service. I would be willing to pay a premium for that.' Here experiences are created within the product or service sold. We embed the 'true' customer viewpoint using VoC data, immersion, the use of behavioural psychology, journey mapping, perception metrics and outside–in design thinking. Here we are starting to think beyond loss aversion and fix, and starting to put noticeable customer value creation at the centre of operations.

Hence, we focus on service and product excellence where subjective experiences are remembered and customers accept a margin on top of the price to experience them. However, this is not full-blown customer experience yet because experiences are derived from the old service and product quality concept: SERVQUAL, something that dates back to the 1980s!

> The SERVQUAL (service quality) highlights the main components of high quality service. These are collapsed into five factors – reliability, assurance, tangibles, empathy and responsiveness – that create the acronym RATER.
> Source: Wikipedia

Examples include:

- How Overbury, a construction company, offered their Perfect Delivery service to the market. Traditionally construction was a low-margin business with returns gained from an onsite confrontational environment. Overbury changed the experience through service excellence, offering delivery 'on time and on budget'. This was most definitely personal and memorable.
- How one Middle East operator identified a key pain point: the excessive queues Emiratis face when they pay their bills at a store on a Saturday. This led to service redesign using tablet payment terminals.
- How a global shipping line and telecommunications operator delivered more service excellence by proactive notification. In the former when a ship was delayed, the company would deliberately inform the customers so they could prepare for the delay. In the latter, they informed customers when a base station was to be closed down. This gave customers time to prepare for the outage and not be surprised when they could not get a signal.
- How Emirates Airlines designs in experiences such as the way the staff speak to you and the fact that the entertainment system is so well stocked and economy food is surprisingly decent. These things are noticeable and meaningful, important when other aspects such as price and flight times are commoditised. They also understand how memorable the human factor is: pleasant staff are more influential than a flashy website.
- How Ryanair has recently focused on excellence but still within their offer of cheap flights. This has led to a 66% rise in sales.
- How Royal Mail has changed the nature of the retail 'experience' they offer, at least in some of their main stores. Now you are greeted, 'hotel-like', by a concierge who directs you on how to collect the right forms, something that can be all too confusing. The environment is also much more open-plan and engaging. They have, at least for now, created an excellent service environment, if not quite a full-on 'experience'.

- Service excellence also considers how different company experiences interact. Hence even though MnS outsourced their call centres to HSBC, they still reach the quality expectations of the brand.
- Ideally, they also 'take' from other industries. Hence, one company created an onsite concierge service based on the Mandarin Oriental hotel, although by creating an immersive environment this actually went beyond excellence as we show.
- And it is about creating the right 'mood', often at a subconscious level. For instance, hotels such as the Leela in Gurgaon, India, design in noticeable and engaging moments. This includes the open and relaxing design of the entrance with classical Indian music playing and elegant water channels weaving their way around the hotel foyer. Think Leela, that memory comes to mind.

Table 3.2 shows how the customer gains personally from product and service excellence.

Table 3.2 Summary of personal gains from excellence

Company	Product/ services sold	Excellence	Personal gain
Global shipping	We ship goods	We tell you when your goods will be late	There is less cost of doing business when I know there are delivery issues
Overbury	We refurbish offices	We refurbish on time and on budget	The way of doing business is less confrontational
Telco	We sell voice and data	We tell you when you can't get your voice or data	I am not surprised and will not need to phone up a call centre to complain
Middle East operator	We sell voice and data	You don't need to queue long to pay your bill	It is less painful to pay my bill
Leela Hotel	We sell luxury hotel rooms	We offer a relaxing environment when you arrive	I feel comfortable and relaxed on arriving after a stressful day

Why Go Further?

Because we are talking about personal and memorable experiences here (which makes it different from loss aversion), there is clearly a close connection between excellence and what is viewed, at least by marketing and brand, as customer experience. So why do we need to go further?

Well the key difference is that in excellence we are still talking about the current goods and services sold; we are not talking about innovating the current goods and service sold based on a deeper understanding of what drives the customer.

Or offering a distinct 'experience' beyond the current goods and services sold: as in create an experience.

And perhaps a little worryingly, the talk can still be quite defensive: we reduce Opex. Of course this is not an issue for the CFO or CEO, but misses the intent of CX and is another trap. Opex reduction is not a relationship KPI, it is a firm efficiency one. Although admittedly this can also be a positive side benefit of CX: so long as CX comes first one.

My argument is, that just focusing on optimising efficiency may be a worthy objective in a static market where customers do not change providers at a click of a button, technological innovation is not disruptive, the social environment plays little role in decision-making and customers are brand loyal. But is that really the world of 2016 and beyond?

Customer Differentiator: 'Offer Something New to Me' Based on My Deeper Drives

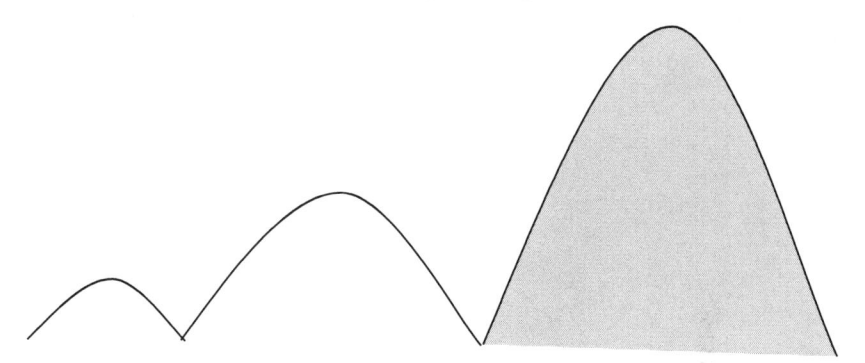

Fig. 3.4 SAM Tier 3. Product and service drives

As service and product excellence become commoditised, SAM focuses on adding positive value and securing higher margins through creating experiences based on how we understand what drives customers at a deeper level. This can arise by innovating the current product or service offer by defining new experiences (Type 1) or offering something distinct from the products and services sold, selling the 'experience' of it all (Type 2).[3]

Customers would say, 'Wow, that's a great new experience.' They would pay a significant premium for the new story over and above the previous goods and service sold.[4]

So here we are no longer talking about selling services or goods. Instead we realise that customers are buying experiences. By way of example if we look at mobile phone purchases, a customer used to buy the network connection, then they bought the services consumed on the phone such as web browsing and now they are seen to buy the experience: 'seeing my family on FaceTime' or 'downloading a video of a football match'.

This in turn opens up new avenues of opportunity and segmentation.

Type 1: Drives related to the current products and services sold

Here new experience-based USPs are found within the current goods and services sold. These target customer's deeper drives and go beyond SERVQUAL. This is what the Cranfield School of Management calls Experience Quality;

although to be honest I like to refer to this as the customer journey quality (CJQ) because 'Experience' for me relates to the holistic of price, service, product and customer journey based on drives (see Type 2).

> Experience quality comprises brand image and communication; emotion, aesthetics, relationship across transactions; peer to peer relationship, social identity and usage process (the journey approach but done in such a way that it evokes something more than the current offer).
> Source: Cranfield School of Management

For instance, how a new Apple Smartphone might evoke an experience that customers are willing to spend money for based on more appealing design and engaging functionality. Sure, it's still a mobile phone but aesthetic appeal is the added value that goes beyond product excellence. At least for now!

This use of aesthetics hence opened up a new market. In fact, Steve Jobs mentioned the potential for this in other areas such as hospital equipment design, as yet unrealised.

Here are some more examples of deeper customer journey innovations taken from the work of the Cranfield School of Management:

- BIC took their brand value for disposability and successfully moved into selling related consumer products such as lighters and razors.
- Rolls-Royce moved from selling engines to selling what their customers really want, the use of the engine or 'power by the hour'. Now, because their revenues are dependent on successfully running engines, they provide engineers at every airport to ensure there are no problems.
- Xerox used to sell printers; now they sell document management systems because their customers don't just want the functional product; they want help with their document needs.

The only issue of course is that this can be a temporary phenomenon. There is only so much time between an innovation being innovative and memorable and it becoming *de rigueur* and unmemorable.

Hence, to innovate around customer drives means you must continuously have a good feel for the pulse of the customer.

Without that the company will risk focusing only on sales today at the expense of sales tomorrow.

Type 2: 'Experiential' drives that create a new market space selling 'the experience'

However, there is another form of experience, one that treats our current goods and services as the platform for offering a whole new 'experiential world of…' realm, more immersive, more motivating and maximising of memory assets. More focused on personal and memorable value than Type 1 by offering something else beyond what we sell already and making it 'The experience'.

Hence, it creates a new market space by focusing on 'the whole experience': we look at this in more detail later on, but for me this is the ExQ in the CX equation ExQ = f(PQ, SPQ, CJQ). Where ExQ is experience quality, PQ (attributes of price quality), SPQ (attributes of service and product quality) and CJQ (attributes of customer journey quality).

Here we revolutionise ExQ which revolutionises PQ, SPQ and CJQ. In Type 1, we just focused on CJQ.

Typically this is seen in examples such as LUSH that takes the concept of 'buying soap', applies the lens, 'How can we create a new experiential world of … soap' and comes up with a new *experience* sale beyond service delivery requiring a focus on design and dramatic quality. Hence, LUSH sells the soap experience! Wal-Mart and Marks and Spencer do not; they just deliver it.

This Type 2 way is best described by Joe Pine:

> When most people use the term customer experience they don't really mean what I mean. They talk about making transactions nice and easy and convenient, what I'm talking about is really a distinct economic offering where you go beyond the goods and services to reaching inside of people, engaging them and creating that experience within them.
> Source: Joe Pine (see Footnote 4)

In this way the customer engages in 'the value of time well spent as opposed to the value of time well saved' (an efficiency brand concept), which translates into more spend more use and more relationship.

Here are some examples, a few of which we go into in more detail, of how Type 2 companies engage at a deeper and more personal level based on what drives consumers.

- We sell soap; what would we look like if we sold an experiential world around soap? The Lush customer would say: 'Let's go and see the soaps', then, 'I need a bar of soap'.
- We sell chocolate; what would we look like if we sold an experiential world around chocolate? The Hotel Chocolat customer would say: 'Let's go and see the chocolate', then, 'I need a bar of chocolate'.
- We sell PCs and phones; what would we look like if we sold an experiential world around our IT products? An Apple customer would say: 'Their products are not just about the technology but about innovation as are their stores and their culture.'
- We sell airline flights; what would we look like if we sold an experiential world around flying? The Emirates customer would say: 'I like the fine touches; the fact I can use Wi-Fi 10 minutes into a flight, they have a great selection of films, the food is good and the staff always make way for me in the aisle' first before they would talk about the flight price or convenience.
- We sell orchestra concerts; what would we look like if we sold an experiential world around concerts? The orchestra goer would say 'I like the fact that the performers get to socialise with us after the concert'.
- We sell online; what would we look like if we sold an experiential world around our Web services? A Zappos customer would say, 'I like phoning their call centre; they are so engaging'. This caring environment is the separate sale from the delivery of a call centre service.
- We sell cars; what would we look like if we sold an experiential world around our cars or even at a higher level around mobility? A future car company customer might say, 'I like their cars; other cars inform me of black ice ahead'. The information delivered to customers is a separate sale from the delivery of a car.

Notice that these companies do not explicitly charge an entry for each experiential world. But there is an important feedback loop that is often missed by companies. When you focus on selling the experience, you create a new condition, a new ecological space for your existing goods and service to occupy! This means that Hotel Chocolat had to completely revolutionise their chocolate products (chocolate slabs anyone!) in a way they could not have done by competing traditionally.

Indeed other companies in this market cannot easily provide chocolate slabs in a normal retail store, because the product is effectively tied to the employee culture, the ambience, the moment, the experience!

So in this way, through CX, they have created a new revenue opportunity not seen before, and are able to charge a higher margin....

In addition, we also see a cultural dependency. In CX, employees become more engaged with the brand because they now cocreate *experience value* alongside customers; hence, we have to invest in their knowledge, expertise and artisanship.

To finish off then, let me highlight some best practice examples of the Type 2 experience, something I call the Big E, which is profoundly based on designing a new and differential market space when faced with commodity pressures.

Best Practice Big E

Here we don't take a *little e* 'sum of everything' approach. The experience the customer has is not about measuring and monitoring everything. Here, experience is used in the subjective sense of 'that is an *experience*'. Something that is immersive has dramatic theatre, based less on product or service design, than experience design.

Rather like the way Geek Squad design in memorable clues that say these are no ordinary technicians! These are Geeks – black ties, Geek mobiles, the Geek badge, the Geek approach – who of course do a stellar job at mending your PC but also provide something else, a branded memorable experience.

Or Zappos, they have a big E approach, designing their call centres to be personal and memorable experiences, celebrating when employees spend more time engaging with customers rather than less time getting you off the phone!

And then there is Starbucks. Here the company created a whole new market by going beyond just selling coffee to selling 'the third place' experience. Customers now go to Starbucks not so much to get a coffee, as to experience the ambience, to meet friends and engage in business conversation.

Indeed, Starbucks has become the poster child for customer experience economics, an often-quoted example of how companies can gain revenue by selling a personal and memorable experience over selling a good or service. It seems there is a premium to be had by creating an 'experience' because you start to 'charge for the time customers spend with the brand' rather than say just for performing the service of giving you a cup of coffee.

Indeed, this is one of the key economic justifications for companies to look at customer experience, frequently used by consultancies and analysts.

The transformation power of big E experience has also been seen in more mundane industries. For instance, Cerritos Library in the United States turned a staid library into an Entertainment Zone. Using experience as a point of differentiation, Cerritos achieved a dramatic rise in membership and created a new market: malls wanting a small Cerritos Library as a magnet tenant.

If Disney did libraries they would probably look like Cerritos.

Mears also use big E customer experience in the mundane B2B local authority home refurbishment business. As a new entrant they faced an uphill struggle trying to compete with incumbents who already had deep relationships with local authority buyers. Of course any incumbent had to compete on price and service but in addition Mears went a step further.

They looked at what happened to homeowners during the refurbishment. How elderly tenants had to leave their homes whilst contractors got to work. Mears therefore decided to provide a mobile home for each tenant, a place where they could rest while the work was carried out. Homeowners loved it. They started to appreciate the experience and provided great feedback to the local authority. Within five years Mears had substantially increased their market share! What's more their competitors started to offer the same; hence the costs of mobile home provision became shared.

In my own work I have seen the financial success and potential of experience used in this way.

With a large construction company, we focused on the employees' experience: the project managers, quantity surveyors and mechanical and electrical engineers. Here value was added by creating a great onsite experience: putting in place 'concierge' services, having great sandwich lunches and so forth. The idea was that because these intermediaries enjoyed working with the company they would be more likely to suggest them as a supplier.

As related to me by Joe Pine, perhaps the big E customer experience can be best summed up in the service industry by the Pike Place Fish Market in Seattle:

In the 1980s faced with bankruptcy the owners decided to become 'world famous' by introducing enticing and memorable moments such as 'throw the fish', where customer's orders would literally be thrown across the market. Sure they still sell fish, but the service is the mere stage for what they really sell: the customer experience. It is this that saved the company financially and made it a world-famous tourist attraction.

And, of course it should not be forgotten that experience also applies to products.

Table 3.3 shows how these experiences have grown their respective firms by providing something different from the goods and services sold, that it is a personal/memorable event.

So returns on customer experience can be achieved by creating experiences that are entertaining, educational, relaxing, aesthetic or escapist; in each case, the firm designs a compelling experience that represents something distinct from the goods and services sold.

And notice how in all these examples, we are differentiating on a drive that customers would not normally express; In Table 3.3 I highlight in bold some of the output senses this refers to. Hence, the process of commoditisation is driving differentiation in the *experience*.

Once firms know what this is, they can then design the supporting stage, whether that is a website, CRM system, shop design, or staff bonus structure, after all:

There is no point in building a stage unless you have a play to perform!

Moving to Transformations

So, in summary, return on customer experience here means more than service excellence or efficiency.

Indeed, to highlight the distinction further, in one of my early articles I characterised how selling an experience differs from selling a product or

Table 3.3 Summary of customer experience approaches

Company	Product/ services sold	Experience upsell	Personal gain
Geek Squad	We mend PCs	Come and see our techno geeks	Memorable moments, a piece of **theatre** that turns a boring experience into something entertaining
Overbury	We refurbish offices	We sell employee comfort	Employees like the firm as they provide the best site to work on: nonconfrontational, great sandwiches, concierge services, a **pleasure**!
Cerritos	We provide books	We sell entertainment	Members want to see the themed zones; they get **entertained**
Mears	We refurbish homes	We sell customer comfort	Customers like Mears as they get to experience a **home from home**!
Pike Place	We sell fish	Come and see the fishmongers 'throw the fish'	Customers love to see the **entertainment** and will pay just to see the display
McDonald's	We sell burgers	We sell the drive-in experience	Customers want to see the new, but temporary, **fun** experience

service. This I called the 'Customer Talk Test', which incidentally shows that the customer experience can itself be commoditised:

When McDonald's introduced drive-ins, customers would say:

> First: 'Wow come and see these drive-ins.'
> Second: 'Oh! And we can get a burger.'
> Hence, it is a customer experience. After a few years, though, drive-ins ceased to be an attraction; customers would now say:
> First: 'We can get a burger at McDonald's.'
> Second: 'Oh! And we can use their drive-ins.'
> Hence, it is not customer experience; it is part of the service.

So experience is not the end! Indeed Joe Pine talks of one example of experience commoditisation, Theme Restaurant Disease, where restaurants tick the experience box with a plastic theme while at the same time producing overpriced and poor quality food at a premium!

Hence we must continue to look to innovate, move with the market and seek new ways to add value to the customer's engagement with us by using better datasets that capture how the experience changes and enables creativity.

Ultimately, though, Joe sees us moving beyond experience towards transformations: where firms get paid on the achievement of some outcome such as paying for those diet meals when we have reached our required loss of weight. And in some ways I can see that coming. For instance, with information quality and the Internet of Things, the deluge of data will necessitate a stronger degree of personalisation using AI and human interpretation. Perhaps we will have our own personal avatar who will determine what is best for us, craft a solution then accept payment upon success?

Summary

So, in summary: in both types of experience we are not so much managing the experience as creating an 'experience'.

With Type 1, we innovated WITHIN and FROM our current goods and services sold based on customers' deeper drives. We do more than focus on price, product and service: we innovate the customer journey.

With Type 2, we innovated BEYOND our current goods and services sold based on customers' deeper drives. We focus on the ExQ part of the equation.

And critically to define these experience differentiators and drives we would start with that question: 'How can we make this an experiential world of…'.

Note: I would start with that question first from which I define the output emotions! Emotion words can be far too general and myopic without an associated meaning. Hence trust in a bank can be constructed not just by trust clues that may or may not be relevant but by numerous other actions some of which are about efficiency. In a further chapter, I talk of us needing to talk of emotional *values* rather than emotions per se, that is, what they relate to.

A Better View of SAM

But there is another way of looking at these categories. Rather than consider them discrete, consider them as different aspects of the same experience which can be emphasised and de-emphasised according to customer drives.

So SAM rather than being three separate economic assets might be better conceived as a bendable sheet that flexes and moves continuously subject to the conditions of the market. It is our role then to manage this asset.

So sometimes efficiency is important, sometimes excellence and sometimes deeper drives. When I buy a coffee, efficiency needs to delivered, it needs to be a clean store! Excellence also needs to be there; the coffee quality is good. And of course the store needs to offer that differentiating experience – a place to sit down and talk with friends – which hits a bigger drive within the customer.

Three aspects in one experience and they all intertwine. And change! They are dynamic.

This approach can be seen by the amount of people who use Starbucks not to experience the Third Place but just to get a coffee and go. In addition, we can see how excellence changes; coffee is no longer enough; we need sandwiches because customers are demanding them in other stores, an added value that can be identified through measures such as Share of Wallet and the Wallet Allocation Rule (Tim Keiningham).

And how innovation around customer drives keeps on moving – everyone now has some variation on a Starbucks theme – what else is new! Hence Starbucks are looking at new experiences such as offering more bar-style services.

This also means sometimes Experience brand status is not required. At the moment I am watching a TV ad for a freshener spray. Cue nice pictures of people creating a great 'refreshing experience'. Well, perhaps I just want it to work effortlessly! That may be enough: service assured, be efficient, focus on UX in its narrowest definition. But then again perhaps we could 'differentiate' by new exciting smells or exceptional brand communication. It depends what you want to do and how the market is evolving!

And that's the thing. Commoditisation leads to innovation. We can only ever be 'so' efficient. So walking along the food court at Abu Dhabi

International Airport surrounded by very similar fast food restaurants I head for the Montreaux Jazz Café. Fantastic, get a coffee and listen to jazz. Music is the experience. It's different; I buy. The only problem is they fall short. Old jazz recordings are played on small TVs, with poor quality sound. This isn't a differentiator! Why don't they get one of the staff to take a break and sing some jazz; how cool is that! Or offer a busking point. Old jazz records for sale, jazz-themed drinks, there is a lot that can be done if they live and love the experience.

And listen to their employees who also live and love it!

SAM Is Not a Fixed Asset

Another thing about SAM is, it's important not to think of subjectivity as some sort of asset in the same way you might think of a manufactured good. I mean an NPS score as with any attitudinal or emotional variable is not just fleeting and forgotten but also wouldn't be thought of unless you had asked the question. Hence, it lacks consistency, resilience and is often gifted.

In later chapters I describe the characteristics of subjective assets and what this means in terms of insights, but for now I reference Olaf Hermans to illustrate this lack of durability in SAM:

Olaf on NPS: The idea that customers score NPS more highly which leads to some abstract conclusion and follow-on behaviour is simply not correct. Although of course as a strategic measure NPS correlates to profitability, as does satisfaction, it does not tell us 'why' or what to do.

Olaf on emotion: Stop reducing the customer to a being that floats on a cloud of emotion. Emotions drive choices but do not so much contribute to 'institutes' like loyal relationships.

And the implications of that are quite important if we want to build loyalty:

Olaf on Loyalty: Relationships and durable loyalty are carefully built by people and organizations living them. 'Conversation, mutually beneficiary and gratifying behaviour, perceptions of motivation and sense-making of others' behaviours' towards us are the key drivers of overall positive affect which we indeed need to feel good in a relationship.

> Repurchase, and purchase frequency, are forms of loyalty that have different drivers than 'invested behaviours' by customers.

So, just moving an NPS metric then is not SAM, and believing in the durability of any old experience holds problems.

For instance, the benefit of experience for me is based on defining and delivering to the deeper drives of customers. So Starbucks was successful because it identified a differential value from the occasion of use, something that was already latent in the experience.

Likewise I would argue the same is true of other experiences. So sure, whenever there is an occasion, there is the chance to make things immersive: whether it's Southwest Airlines' singing air stewardesses, Zappos' call centre operatives, or Geek Squad's technicians dressed as geeks.

The problem is that although many times this adds indirect value to the experience the customer has, and is something different from the product and service delivered, unless it hits on a deeper customer drive to use or at least accentuates it, then it will not be durable! Hence, the importance of the *driving* theme! Less so the memory per se!

So Starbucks, the Third Place, is a durable added value.

LUSH's focus on aesthetic appeal and designs of soap is a durable added value.

Geek Squads' technicians dressed as geeks are not durable, but important nonetheless as an indirect modulator of the experience we wish to have: 'The PC is fixed.'

The CX Equation

So far then I have defined customer experience and provided a framework for understanding how returns are to be made Critically there has been a connection to customer value through the CX equation: $ExQ = f(PQ, SPQ, CJQ)$ This comprises:

1. Attributes around price. Value is created by manipulating price sensitivity (PQ).
2. Attributes around product and service quality. Value is created by manipulating SERVQUAL (SPQ) and value-in-exchange.
3. Attributes around the customer journey. Value is created by manipulating customer journey maps (CJQ) and value-in-use.

4. The overall sense gained of the Experience, as a function of each (ExQ) from which we innovate. Value here can be created by market growth and differs from CJQ value-in-use by its focus on the personal and memorable aspects. Hence, CJQ value-in-use means power by the hour from Rolls Royce; ExQ means personal and memorable dimensions that create a new market space such as in Hotel Chocolat.

For me, each one of these dimensions is used to a greater or lesser extent in ExQ design. So LUSH and Hotel Chocolat make the product an ExQ while also revolutionising PQ and CJQ. Emirates Airlines makes CJQ an ExQ while maintaining PQ and revolutionising SPQ. Poundland makes PQ an ExQ while altering the CJQ to reflect this. Zappos revolutionises the SPQ while wrapping in changes to the PQ and CJQ.

But always remember, CX flexes. So, digitalisation can makes things cheaper which for a while engages PQ (as in Uber); or makes things complex which for a while engages SQ as a competitive differentiator. Or things that were experiences become degraded as in Theme Restaurant Disease. So long as we see the whole equation – the EQ – we will be aware. We become myopic if we only see its parts.

Information Quality (InfQ)

In addition though, as experiences become commonplace, I would say that firms are starting to look at a new source of value: information quality, where InfQ sees consumers driven by the impact of the information they receive and use.

For instance, how companies interact with them and how customers use information to interact socially between themselves. In other words, they are competing on the arrow part of the DEM model!

And information can be about process efficiency, cost saving and Opex reduction or it can move customers.

For instance, airlines, hotels and restaurants use information about you to deliver more personalised experience; car firms share information between vehicles about road conditions so that drivers know when there is 'ice on the road' ahead; Amazon uses your data trail to provide relevant offers and suggestions for next best actions, E.ON uses information to advise customers on how to save money on electricity and gas, and the use of health care information from Google Analytics will allow us to prevent ill health.

Information quality can therefore be seen as a platform in its own right to create personal and memorable 'interaction' experiences in the same way as goods and services. Just look at Watson AI.

And this is happening now! Journeys are being disrupted and redesigned on the basis of pushing next best messages that are emotionally resonant, that are personal to the customer.

But what is the ROI of all this?

Well, consider this. Not only is this about value creation but InfQ creates moments of use, and the more you engage with the provider the more you are likely to stick with them for the long term. As long as, of course, it's the right sort of engagement! The type that helps build a relationship, from which your NPS, CSAT and emotion scores arise.

And one way to reinforce this effect is through the *effective* design of use moments. Creating disruptive journeys for good not bad interaction, using that to create a future impulse to buy. Or how about in the design of services and products enabling the customer through information to self-select how use is configured rather than how the firm would like it to be configured: less a company's and more a customer's invitation to treat.

If everything is reconceptualised as information, the opportunities are endless.

Hence InfQ is about:

Data Sharing and Analytics

For instance, sharing information streams to come up with customer experiences, as we saw in the case of ice on the road, or correlating objective and subjective information: now a pizzeria could determine at what temperature pizzas correlate with 'good performance' in the mind of the customer. Hence they could then put in place a relevant process to maintain pizzas at certain temperatures, or effective use of data analytics on website use 'with the customer experience in mind'.

Or how about after getting a puncture, my insurance firm receives a message from the RAC and can send me advice that my insurance will not be raised. Critically the message is sent while the emotion is fresh, during the repair or very soon afterwards.

Or sharing information across departments: using a single view of the customer to help break silo behaviour.

With the coming Internet of Things, there will be myriad ways information could be turned into products and services that are meaningful to the consumer, for instance, the ability to turn your heating on from a mobile phone app. The capability to augment how you live your life and do your work: remember Watson AI that is already having an effect, enabling cancer doctors to reach a quicker consensus on patient treatment through its capability to 'fast sift' through research data and serve up possible scenarios.

Likewise in the pharmaceutical industry we see how we could, with the right information on use, design disruptive customer journeys. So, doctors could up- or down-dose medicines dependent on the individual patient, as they use their drugs.

And there is the whole concept of defining next best actions as we have seen.

Hence, to effect change requires working cross-silo and *cross-company* and frequently proactively. After all when the world is just seen as data, everything is a commodity! And this isn't as far-fetched as it sounds: 3D printing, construction designs changed at the push of a button with associated guarantees in price, virtualisation of networks; everything is not an experience but information! Experience comes from how we engage consumers subjectively

Which means the only difference will be in how we share, collaborate and create new mash-ups between companies that define new sources of symbiotic value and engage our drives beyond the functional.

InfQ it seems has the capability to deconstruct the firm in the same way that ExQ has tried to deconstruct the silos.

Who manages the human parameter best will win!

Being Social

Information sharing offers the opportunity to create social communities and microsegments, as we saw with Golfing Gertie, or as a platform for crowdsourcing and ideation. As well as offering the potential to engage in social dialogue with the consumer – dialectic surveys – that ask relevant questions to support the relationship. Or the provision of information that augments human knowledge to support the encounter.

Self-Expression

In InfQ the customers define the offer and the journey that's best for them derived from the offers we can provide!

For instance, take personas: why don't you spend a little bit more time deriving psychological insight into how you create them then serve them up as SELF-SELECT options to the customer because self-serve best takes account of the subjective.

So shopping in Morrison's, today I will be 'target shopper Steve' and the next day I will set things to 'kids in tow Steve'. That way I will allow you to send me interactions that are relevant not abstract.

Or how about applying story metrics that are quantified by the customer themselves, not an algorithm, as we show. These will then allow the customers to define what is relevant to them.

Early Stage InfQ

This information quality trend is very much at its early stages and still operates mostly at an efficiency level, as we saw with the Big Bellies. But there is no reason why this should not change. However, I have to say that in my experience so far things have not been good. In fact, it seems as if we are in the midst of a period of increased complexity for customers and employees due to IT and a drive to make departments less relationship driven and more functional because in their view of the world the functional is measurable but value is not! Even if it's value that matters most!

For instance, driven by Opex reduction many major banks are going online. The effects of this are, however, not entirely good for customers. For instance, I recently wanted to put a cheque into my account. This meant opening an instant saver account.

Unfortunately what used to be easy is now a nightmare.

Firstly, a date of birth error was found on my account. Now I would have expected this to be rectified by a phone call, but no, I had to go in person to the branch for a 'personal appointment', which actually meant, because branches are now under pressure to sell as everything is online, a barrage of questions about how I should get some new reward account.

Sorry, not interested; you have wasted my time; can we just get on with it. And of course the person doing the selling was particularly rude, unsurprising because all that counts is sales not the relationship.

Then I find 20 days later, my account still had not been opened. Cue a phone call to their call centre that first of all said I had received a letter 20 days ago about the need for my mobile phone number, when in fact that was untrue because the letter was about asking me to go into the office to change my date of birth! Which incidentally had been incorrect on their accounts for 30 years; no doubt they blamed me for their clerical error.

Then we finally get to the bottom of the matter. The original office appointment to change my date of birth had in fact not been completed. The office had not sent my details by email to the relevant department who would then have no doubt reinput the details to complete my application. All this of course, I had not been told about and would take another week to resolve!

In such ways we can see how IT complex processing and multiple systems and compliances have in fact destroyed the experience.

And at the end of the day, it was my great interaction with a person in a call centre that saved this experience: the humanic is more memorable and influential!

Technology needs human interpretation to move from efficiency towards experience (Table 3.4).

Management Implications

1. There is always a customer experience.
2. What's the ROI of customer experience is like the question, 'What's the ROI of innovation': the real question is: 'What's the ROI of the various ways of doing customer experience?'

Table 3.4 Quality questions

Differentiator	Quality	Company offer	Example	Company actions
What we buy	Product	Price Product features	£1 Cup of coffee	Sell it
How we buy	Service	Great service	£1.50 Sit-down restaurant	Create a buying environment
How we experience	Experience	An experience	£2 Peruvian Coffee Alcohol £2.50 A place to work	Create a personal and memorable environment
How we receive information	Information	Relevant information	More regular use, journey disruption personalised by me	Create a sharing environment

3. The core principle is: how we add value to the customer's account with us not how we can add value to our business account with the customer!
4. The paradox of customer experience is: we only ever see the benefit of our CX strategy after we do it; and if we want to predict the benefit beforehand we will only ever end up engaging predominantly in efficiency.
5. The return on the various ways of doing customer experience can be divided into three areas: efficiency–excellence–drives: subjective asset management (SAM), each exhibiting a higher rate of return.
6. UX is supposed to be about a CX focus in product and service design. The reality is it has been degraded into a term for 'make it work.'
7. Efficiency is the platform: CX is the relationship.
8. SAM is not a linear process but exists in the one experience all the time.
9. SAM is not a fixed asset: it frequently lacks durability. Hence the focus on drives over memory.
10. There is a CX equation: $ExQ = f(PQ, SPQ, CJQ)$, where the function relates to drives (goals and subgoals).
11. Information quality defines a new source of value.
12. Information quality (InfQ) is still at an early stage.

Notes

1. Of course we can see drives strike across all economic areas, but it's the deliberateness of the approach that is important. Assurance seeks to assure, Excellence seeks to excel, Drives seeks to look deeper at what is beyond the current offer.
2. Fix is often conflated with process efficiency and cost-cutting, but you can merge the two, so customer experience is a safety valve for Opex reduction, ensuring that any cost-cutting does not chop out value and acts as a safety valve for sales growth, so that sales growth leaves the experience at least unchanged. But in an ideal world CE is your lever of growth.

    In addition, a focus on hygienic process efficiency can release time to engage the customer so even here they are not mutually exclusive. The retail firm that 'Apple like' gets their shop assistants to take payments via tablets means they also have more time to talk to customers and engage them in the products on sale. As long as other customers are waited on and don't see a greeter standing around!.
3. I call these Type 1 and Type 2 rather than put them into distinct categories because they are both about deeper engagement and drive. Academics call this value-in-use.
4. The paper, 'Network as an Experience Platform' (Source: Pine, Walden, McCann-Murphy, 2013) emphasises this point.

Part II

Data

In this section, we look at how customer experience data being subjective requires a different level of understanding. If we don't do this, we will end up treating subjective data as objective and end up creating an Efficiency brand not an Experience brand.

4

Right Data

For something to be a customer experience it must be experienced by the customer! In other words, experience is subjective. A key question for managers then is, 'If we are managing subjective experiences, how do we measure them?'

Well, to answer that question we need to understand how subjective experiences behave, how they are different from the type of objective data we are used to dealing with.

Objective Data

Objective experience is anything that is tangible, for instance, download speed, aircraft wings, and radiators. Because these things exist in their own right, independent of any human observer, they have a fixed meaning. Hence, download speed is download speed, aircraft wings are aircraft wings and radiators are radiators whoever is looking at them.

Likewise, any interactions between objective experiences (let's say the properties bolt tightening and seal tension) are mathematically fixed. So, a loose bolt on a radiator will lead to a water leak in a predictable fashion:

cause and effect are clear and a fix (tightening the bolt) is easy to execute with predictable effect: that is, no more water leak. In effect we can say that nature is the measure of all things and that measure is called mathematics. Because observers are not part of the system, there is only one truth and a mathematically definable meaning.

These experiences are ideal for a command and control style of management because you are able to state what more or less of a quantity means (more tightening, tighter seal), know how it predicts effects and hence know what you should do about it.

There is a root cause!

And because, at its heart, objective data are quantitative, the 0s and 1s of mathematics work and predict outcomes. In this environment, an engineering, linear mindset focused on process efficiency, regression models, Lean and Six Sigma work well. We reduce waste in the system as it represents not creative iteration but a failure to define root cause.

Subjective Data

By contrast, subjective experience is anything that relates to a physical something but is itself intangible; for instance, 'look and feel of', 'ease of', 'friendliness of', 'satisfaction with or recommendation towards' a website.

Because these experiences only exist in the mind of the observer (the unit of measurement is the human brain), they are not fixed in meaning or importance. Hence, if we take the example, 'look and feel of website', we will find that 'look and feel' can take a multitude of meanings and be open to reinvention in a way that objective experiences are not.

A bolt is a bolt; look and feel can take many directions.[1]

In effect we can say that humans are the measure of all things and because observers define the system, there are multiple truths and fuzzy meanings.

The implications of this for managers are that if they end up dealing with any human 'subjective' system they will end up managing fuzzy data. And in this environment we cannot act in the same way as if we were dealing with machine data. In short, this is an open not a closed system.

Hence managers have to contend with the fact that consumers are not cost-benefit calculators, and instead pull memories from their minds in anticipation of an event and in the moment, memories that interact with each other as neuronal connections in the brain at the moment a decision is made, memories that are not fixed through time but only become durable if they are meaningful to me.

So, whereas in objective experience everything is quantitative and driven by root cause, in subjective experience things are *qualitative–quantitative*, more emergent and interconnected. Some things are root cause but some things are not, being instead frequently indirectly influential on a decision to buy or browse. Customers also frequently do not so much evaluate experiences as they receive them as describe them as salient or not against relatively abstract criteria such as 'does this company care'.

To illustrate some of these effects we can see how a supposed direct effect on behaviour such as 'I buy this brand of bread' comes with indirect drives such as look and feel of packaging and also things outside the immediate purchase such as ease of parking, friendliness of staff and so forth. Of course, it's more difficult to say what the ROI of friendliness of staff is, but it is a drive nonetheless, frequently nonconscious in expression and when emotionally impactful can knock out a core need! Cool packaging! I'll spend 10% more. It's difficult to park, I'll go elsewhere.

The Importance of Modulators

In summary, subjective experience also engages changeable indirect effects known as modulators; they represent how the experience the customer has is made up of both base needs and sometimes small fleeting moments that interact with each other to change how we perceive things. Together these make for a complex system.

Here is a paragraph then from a public blog that references Dave Snowden's thinking on modulators (one I recommend people read as it summarises some key thoughts very well):

> The following metaphor is useful to understand what this (modulators) means: Imagine a metal table with several powerful magnets spaced around underneath. On top of the table are iron ball bearings. While the strength

and polarity of the magnets remain constant, the bearings form a stable pattern on top of the table. If one magnet changes polarity and the others remain constant, the pattern formed by the bearings will change in a predictable way. That would be an example of a driver in systems thinking terms. In a complex adaptive system however, when one magnet changes, the others change as well in ways that don't repeat. The patterns formed by the bearings therefore become unpredictable, depending on which of the magnets have changed and in what way. The magnets are examples of modulators, factors or forces that influence the system at the same time and in unpredictable ways.

https://sonjablignaut.wordpress.com/2013/10/28/5-differences-between-complexity-systems-thinking/comment-page-1/#comment-439

Hence we find aspects of experience that are changeable, colour our experience, affect how we behave and how we create a sense of satisfaction, ease of doing business and so forth, yet, of course, work in conjunction with more root-cause drives.

And I can demonstrate an example of this in my mobile phone buying behaviour.

Walking into a Phones4U store a few years ago, I was struck by the intense discount merchandising, the loud colours and the sharp-suited sales-driven staff. As soon as I walked in and expressed an interest I was asked to sit down, shown a contract and the cheapest mobile phone in town.

Now normally I would be attracted to a cheap price and a good mobile, but these other aspects of the experience said something else to me.

And what do you think I did?

Walked out of course, went to the more expensive but more appealing O2 store and bought from there. Price for me was a root-cause drive affected by the modulator 'experience'. Hence, the firm's assumption that the former was all that mattered was a very naïve view and one that damaged my custom: because the stress was on sales targeting and delivering a probably unintended poor customer experience.

Focusing on root cause alone is therefore not wrong but constraining and blinds business to only those customers who would sacrifice the experience for the discount. Hence it is essential to be open to more contextual and emergent factors affected by modulators (peripheral clues), emotional value and how the market is shifting; after all these factors develop a sense of 'what it feels like to shop here' and 'why', things that also open up creative potential.

And remember even if you changed the price, there is no guarantee of a straightforward predictable effect because it depends on other factors: what our competitors are doing, how it's executed, and so forth.

Buying Bread

To show this further, let's go back to the buying a loaf of bread example.

Of course there are some root-cause effects, the price is right, the product features are right; that's why I go! But it could be argued that the price (£1.45 for that loaf of bread) or the product features (wholemeal) are commoditised or enough already.

I for one regularly buy the same type of loaf, Warburton's. I hardly look at the price, it's what I need, and it's what I'm used to. I am attuned to the style of packaging and its position in the store. If you were to ask me emotionally how I felt, it would be pretty neutral. I don't need to feel good; it's an intrinsically standard 8 out of 10 kind of thing. But if you looked at my purchase behaviour I would be tremendously loyal because I am habitualised: which for me means I have a strong nonconscious attachment to the brand and am inert to looking elsewhere; I am perhaps more 'committed' than loyal (note the difference to normal views that loyalty is all. For me, at least, what may be more important is holding resilience to change and an emotional norm to the product or service in question).

So I am unlikely to react at the margins to a price rise. That is until my partner says they are ripping us off. Likewise, I am desensitised to the product features. Again, as long as it does what we are used to and there is no change in the market for bread styles, I am habitualised to Warburton's and I buy what I'm familiar with, at least for this category.

However, as a manager I could move beyond competing on price and product features and start to innovate around the deeper customer drives associated with indirectly (or hidden, directly) important things. I can start to create the bread-buying experience.

For instance I could look for new value dimensions. I could release products that compete on how we *use* the product, the 'toastie', 'the speciality healthy loaf' or compete on *aesthetic* features such as attractive packaging design, getting hold of our in the moment curiosity. Or I could extend into other areas related to bread purchase such as Warburton's cooking equipment.

Alternatively I could focus my efforts not on the product (bread) but on making the environment attractive to footfall; such as opening a speciality Austrian in-store bread bakery; creating an 'experience' not just managing the experience. Likewise, similar to Domino's Pizza, I could make visits to local schools to show how bread is made or like Metro Bank, where school trips go to the bank vaults, I could encourage visits to a Warburton's bakery.

And how about Warburton's interaction with the shop itself; are there any B2B activities that could create more of a relationship?

Or perhaps if we look from the point of view of the store, could a manager here make the more general bread-buying experience better; perhaps by focusing on the whole customer journey, for instance, friendliness of staff, ease of parking and so forth; all indirectly impactful on buying a loaf of bread, but influential none the less.

So next time you go to the supermarket, ask yourself the questions: what is the ROI of the music playing in the background, the ambience, the friendliness of the customer service reps, the ease of finding what you want, the ability to find a parking space, the wait time at the till and the way they interact with you. It's not so easy to put a figure on these things individually, but consider what the environment would be like without them!

Price and product are not the only things that drive or could drive ROI. And even these are subject to subjective interpretation!

Hence we need and must understand subjectivity to do customer experience. And because managers are so used to looking at data as objective it is important for us to understand the difference in how subjective data behave, lest we treat subjective data as objective and end up with a grey experience competing only on price and product features! Or treat the intangible as tangible and end up buying an IT system.

Of course I know what you are going to say: if we can find more efficient ways to make bread we could cut the price. Or we could provide it cheaper by using lower quality grains. To which I say, no one said efficiency wasn't a play; it just depends on market conditions and where you want to compete!

Management Implications

1. Objective data are mathematically defined: subjective data are defined in the mind of the customer.
2. Subjective data are fuzzy data.
3. Subjective data are qualitative–quantitative and interconnected: they comprise modulators as well as direct effects on behaviour.
4. Subjective data operate in a different system to objective data.

Notes

1. Sure you get different types of bolt, but it's hoped you appreciate the qualitative difference.

5

Some Key Things That Make Subjective Data Different from Objective

Subjective Experiences Emerge and Modulate

When we look at a subjective experience we find that it comprises a lot of interrelated smaller experiences; for instance, the look and feel of a website relate to many things as we can see below in Fig. 5.1.

This is analogous to objective experiences where say an aircraft wing comprises many smaller components. However, there is a critical difference.

The sense of 'look and feel' doesn't exist other than in the head of a subjective actor – you! Which means it is derived from mental processing. And when we determine the quality of something through mental processing we are frequently not engaging in summing up all its components in a root-cause mathematical way.

Instead, we derive a subjective sense of 'look and feel of website' as it emerges from the interrelationship between its multiple attributes, from its modulators. Of course the experiences (modulated or otherwise) we pay attention to are also the ones salient to our well-being (our goals and subgoals), as determined from prior expectations and how value dimensions are created in the experience.

By contrast, if we were all human calculators a sense of look and feel of website would not emerge; it would be derived in a mathematical root-cause way and we could answer with confidence the question, 'What's the quantitative impact of a unit of niceness of font on look and feel of website?' Or 'By linking look and feel of website to spend; what is the ROI of niceness of font?' A patently nonsensical question, but without a nice font there is no nice look and feel!

Similarly, what is salient wouldn't come into it. Everything would be pure maths rather than a question of 'When is a tipping point reached?' and a new sense of value created.

Here is another example, where we see how the sense of trust emerges in a bank.

Take Metro Bank. Their consistent and intentional approach to providing an open environment – friendly staff; branded clues and other aspects – doesn't engineer trust in a mathematical granular way; instead trust emerges from our mental processing of all these elements as we engage with the experience. If you like, Metro Bank enables a trust environment that allows 'trust' to flourish.

Not forgetting of course, that trust can also be destroyed by a root-cause action, such as 'They were rude on the phone', and that these change through time!

For me this means that we need to manage both the fixed elements of experience that express brand delivery and the not so fixed elements (the modulators), as well as be aware that a sense of trust or anything else is not a calculated variable from numerous additions (as in 'no pens on chains' + 'warm blue carpet' = trust!).

And taking that approach means we must concentrate more on understanding the behavioural psychology of consumers! But of course, this is something that companies don't like to consider. Indeed, through gaming the data, and ignoring how we frequently describe experiences using narrative companies actively remain myopic to how customers think and feel.

For instance, the IT company that puts in a couple of NPS questions on a survey correlates these to objective technical KPIs and finds out that NPS falls when the network goes down! So far so good, except then the company says, 'Well in normal conditions when the network is working, NPS is higher so let's draw a line between the two states and say that NPS improvements are dependent on network improvements all the way up to 10 out of 10: no confounding factors.'

Efficiency now becomes delight, because you have put the assumptions of a statistical model before the data. And your only inputs are the technical KPIs you want as inputs!

Or how about the company that uses selected statistical assumptions of regression to 'assume' there is no correlation and messiness in the relationship between 'niceness of font', 'ease of reading text', and so forth. All this means is they have forced subjective data to behaviour-like objective, erased any emergent and modulator characteristics and come up with some root-cause nonsense that sits well with chief executives, CFOs and CTOs, and focuses us on efficiency for no reward and leads us to waste huge sums of money on 'hygiene' which keeps vendors fat, rich and happy but does nothing for your customers.

In each case companies are using response bias to 'prove' that we are all just human calculators!

> Objectifying Subjective Data
> Response bias can have a large impact on the validity of a survey. This bias is caused by a number of factors, all relating to the idea that human subjects do not respond passively to stimuli, but rather actively integrate multiple sources of information to generate a response in a given situation. (Wikipedia)
> To put it simply, this means if you stick a questionnaire under the nose of a customer you will always get a result because you asked the question! But that doesn't mean the answer is analogous to reality.[1]
> For instance, I could give you a survey right now. To what extent would you recommend the type of font used in this book? Of course you will give me an answer but that has more to do with the fact I asked you the question than any real concern or thought about 'type of font'.

If you still don't believe me, look at the avalanche of predictive analytics data out there to prove a subjective X leads to a behaviour Y! All different, all not particularly predictive at all! As one research house engaged in churn told me, 'Well we can't give you a solid prediction of decreasing churn by X%, because life changes.' To which I say, 'Well why not embrace the change, and adapt your metrics to it.'

Nonetheless, don't let this sound negative. I accept that up to a point you can get some good predictors, such as you might find with 'look and feel' as a predictor of online satisfaction or NPS. But beneath that abstract descriptive surface, understanding what drives good look and feel or any other such phenomena gets you into a more complex, emergent, modulated and critically dynamic structure.

We need therefore to be more open when dealing with subjective data; we should be more semi-constrained and oblique.[2] But this shouldn't be something we fear!

For the manager it means we cannot expect to predict a result for many experience scenarios involving subjectivity. And nor should we! If we do and fit ourselves to some statistical root-cause model, we constrain ourselves, reduce our understanding of what 'value' is or could be and limit the opportunity for experiences to emerge serendipitously, which means if subjective data were objective we would not have creative potential.

So of course we want to deal with subjective fuzziness, after all without it there would be no Starbucks, Apple or Amazon. We would all be deeply interested in functionality and the world would be a very grey unchanging place indeed.

The great failing of companies in customer experience then is that in their rush to root cause they limit themselves to managing a few negative experiences such as you might find in complaints data and miss the opportunity, the bigger picture, the understanding of how the market is changing and the need to test out new value-added experiences.

Yes, you can find out the ROI, but execute and enable first.

Subjective Experiences Are Fuzzy

Because subjective experiences exist in the mind of the observer they are fuzzy in meaning. Hence the definition of look and feel – what is a unit of look and feel anyway – is influenced by outside influences such as brand advertising on TV or a new invention: hence this is an open not a closed system.

This is unlike objective experience where the 'meaning' of an aircraft wing bolt is fixed and not dependent on the 'meaning' of the aircraft headlights.

Hence, fuzziness means new subjective experiences will surface through time to affect what it means to say *feel care or trust*: as shown by the 'other things' influence on Fig. 5.1. So, in our earlier Jazz Café example, maybe, people start to like it, other cafés see this and respond, offering music as a differentiator or upping their game in other experience ways.

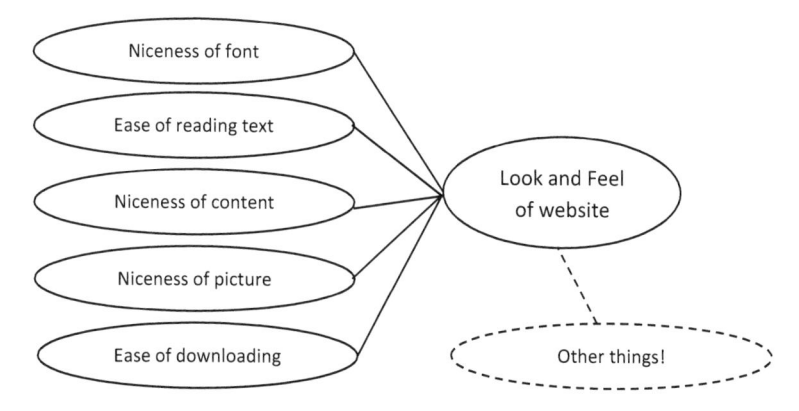

Fig. 5.1 Subjective experiences are fuzzy composites

The system evolves, what it means to like a café experience changes and if you don't evolve with it you will soon find yourself extinct.

Fuzziness also means you can't assume that just because say something like empathy is strongly correlated with say NPS, we should target this alone. Customers for instance may consider empathy enough, 'Yes its important but I don't need any more. In fact I may be more concerned and open to other indirect aspects of the experience that colour my impression of you. Maybe I'm bored with empathy! Maybe technology degrades the importance of empathy as a differentiator.'

And if empathy affects satisfaction/NPS its fuzziness also means things will work the other way round. So NPS brand scores affect empathy! Hence an Apple call centre will get higher satisfaction scores than a Microsoft one, even when the experience of empathy is exactly the same.

(And both NPS and look and feel are really affected by use and familiarity because behaviour drives attitude!)

Once again an efficiency brand will totally miss how things are playing out in this messy fuzzy way because efficiency is inherently myopic.

Indeed, fuzziness is the reason why efficiency methods of research that isolate an item, let's say cleanliness of floor, from its system, let's say grocery shopping, are so flawed. In isolation, customers can be made to compare a clean floor with a less clean one, and state an improvement. But as soon as cleanliness of floor is placed in context with the grocery shopping experience, any isolated data that assume floor cleanliness drives value are quickly overwhelmed.

But this isn't a council of despair. For while fuzziness may cause some problems:

If we can start from the point of view of defining what drives or could drive customers then we will have a better chance of moving the dial in terms of creating experiences of value,[3] which of course means we have to understand customers' personal world and personality, how they interact 'in the decision moment' and construct memory after the effect.

And if we understand modulators, we can focus our efforts on small changes that do not cost so much but are part and parcel of bigger effects. Rather like the Nudge effects described earlier. An example of this could be how I asked one operator in a multi-SIM environment why they couldn't make their SIM more colourful and engaging, encouraging its use.

Also, if we understand that this is an open system, we can look to other experiences in the industry or outside the industry that we can bring in to our experience. This is particularly important because customers frequently sacrifice to what they know already.

So with a travel company targeted on NPS, I identified the core customer drives to book a flight. This led to a redesign of the website where the flight map moved from three clicks down to the front page. Likewise, and in conjunction with this, I identified the need for a better look and feel, particularly apparent when compared to one of their direct competitors. Again this led to a wholesale redesign and even though in both cases the fuzzy nature of 'ease of booking' and 'look and feel's' relationship with NPS meant we could not state an improvement would happen, we could state a belief that these changes would make the experience more motivating.

Small things lead to larger things; it doesn't matter if they directly change an NPS score or not. You measure return over time; you trial and test; which is exactly what we did.

So, in essence, Experience brands understand fuzziness implicitly and seek instead to influence the direction of travel. They respond resiliently to customer issues as they arise but also make experience more resonant in the mind of the consumer. Experience brands say, 'It's motivating; let's do it!' Not 'It's motivating but will it or won't it move NPS!' Experience brands are open to evolving; and with InfQ evolution is now happening at an ever faster pace.

Here's what Dave Snowden of Cognitive Edge has to say in relation to some of these issues:

> As a system becomes more open to multiple interactions among multiple things, every interactions becomes inherently uncertain.
> You miss the opportunities (if you assume traditional driver analytics approaches), if you understand where people sit at the moment, you can influence their direction of travel and you are much more likely to produce a good outcome than if you force them to a particular end-point. The reality is things are constantly changing, you haven't got predictive systems.

Subjective Experiences Are Not Entirely Recallable!

Subjective data is like an iceberg. At the tip of the iceberg are the memorable moments that drive customers; these the customer can recall.

Beneath this are other aspects intrinsic to an experience that gives it a 'sense' of value. So I can't remember a single episode of *Rupert the Bear* but I have a 'sense of its quality'; even if that's from what others say! Likewise, many customers in their recall may just say, 'Its OK', 'Its reliable', 'No different but good.' They build a sense of the experience from moments long forgotten or what is socially acceptable to say.

Of course we shouldn't get over-obsessed with the 'nonconscious' or conditioned nature of things, but it does mean we need to probe this in our research.

So customers, it appears, don't so much lie, as express a number of biases when they answer a survey. For instance, although there is a recallable need: 'I need to buy some car insurance' so 'I phone a call centre' looking for the best price, this need comes with strings attached. If I find a call centre operator who is not so friendly this will affect my 'belief in the price being good' and the type of place I want to do business in. But of course, these impacts are muted when I answer that survey because I post hoc rationalise my decision – it always was a price-based decision!

Alternatively, of course, it could be that a call centre was less friendly but I got the price I wanted; I'll go back! (Emotions and experience can also extract value from the so-called functional.)

This is all very unlike mathematics that assumes customers are capable of knowing all the inputs into how a score is constructed. So a measure of web browsing speed 100% explains the speed of web browsing that your network probe dictates. By contrast, NPS has no such analogue. Which is a good thing; otherwise as I have said there would be no creative potential in the market!

Subjective Experiences Need Context

Unlike physics where a number refers to a quantitative thing such as speed, in subjective data the number refers to a qualitative thing, hence the context, the meaning of the number matters more. For instance, think about how 8 out of 10 for product quality in Aldi is not the same as 8 out of 10 for product quality in Waitrose. This is why I get a little annoyed with the move the metric people! Sure it sounds logical to target linear improvements such as improving NPS by +10 points. But we can also stay the same but change the meaning of a score and drive higher value.

For me this context characteristic of subjective data is rather like figure-ground effects. You need both the figure and the ground, the context, to make sense of the situation (Fig. 5.2).

Ground

Root Cause

e.g., price

Figure

Emergent, modulators

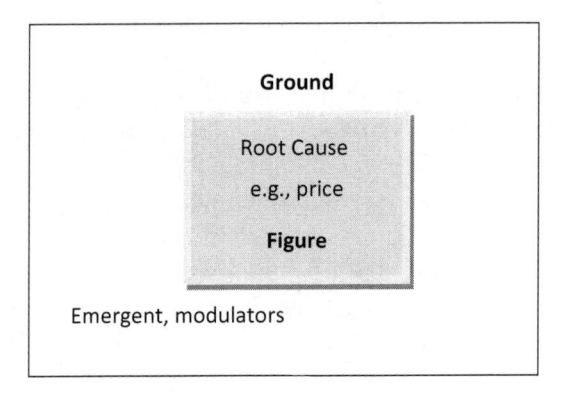

Fig. 5.2 Figure–ground in CX

Figure–ground organization is a type of perceptual grouping which is a vital necessity for recognizing objects through vision. In Gestalt psychology it is known as identifying a figure from the background. For example, you see words on a printed paper as the 'figure' and the white sheet as the 'background'.

(Source Wikipedia)

So in CX *figure* might represent hard root causes that affect the experience, such as price and efficiency. By contrast, the rest of the area, the *ground*, might represent something 'emergent', the 'modulators', the context made up of many small deeply interrelated qualitative experiences. These affect the memory of the experience that customers take away and how they feel towards the root causes as we saw in the O2 experience or how we might say, 'I have a sense of trust.'

And notice how this context makes up most of the space; at least in my opinion.

So the context of trust in Metro Bank, 'innovation' in Apple, 'care' in Emirates Airlines is the *ground*. Something which emerges from multiple aspects, for example, not just in a fixed-price sensitive way but from environmental clues and how we manage the customer–staff encounter.

To go back to the figure–ground analogy this means we must manage CX context effects by becoming aware of their edge-augmentation, which means how shifts at the edge through time influence how we perceive the overall sense of emerging trust, care, innovation and so forth. Hence, we need metrics that are not fixed or rigid but pick up these changing nuances and focus our design.

I'm not sure if this is a good analogy or not for subjectivity; I'm sure many chartered psychologists will have something to say about that, but it gets to the point nonetheless: both root cause and modulators matter, both influence each other, and watch for shifts as these tell us about dispositions.

Subjective Experiences Have an Intrinsic Quality

When I was working with one of the UK's largest electricity and gas suppliers it was noticeable how in focus groups consumers spoke of purchasing their supplies as a 5 out of 10 type of thing. This sounded bad, but from verbatims it meant nothing of the sort, just bland and OK.

This is something companies need to take account of in their understanding of experience. Not all of it requires high scores to be meaningful. Sometimes things are just hygienic and that enough is enough and no more. Not everything needs to be 8, 9 or 10 out of 10 to be motivating. Although of course, some categories such as Fabergé eggs and luxury hotels 9 or 10 out of 10 is most definitely a score you should aim for.

Of course this intrinsic qualitative nature is very different from objective data where speed, weight, tension and numerous other mathematically definable properties are cross-comparable and nonqualitative.

So now we understand that subjectivity is different, what do emergence, modulators, fuzziness, flow, change and context mean for how we set KPIs?

Subjective Dashboards Are Different

An objective experience is fixed in meaning; hence we can say, to increase the speed of a car you press your foot down on the accelerator and you will get more speed (Fig. 5.3).

By contrast a subjective experience does not have a fixed meaning. Hence, if we want to achieve an increased return we should seek not so much to achieve more of an experience but seek to change its meaning.

Fig. 5.3 *More of* objective dashboard

5 Some Key Things That Make Subjective Data Different...

100

8

80

8

40

8

Subjective
Dashboard

Fig. 5.4 *Change meaning* subjective dashboard

Hence, in Fig. 5.4 we can see how although each dashboard scores 8 out of 10, they relate to different experiences, delivering different returns of speed (denoted by the colour change). So we don't want quantitatively more of something. We want it to be qualitatively different in order to change the speed.

Not forgetting of course, that we can, at lower speeds, get an increase from quantitative pressure.

If this was a machine we would say, as you put your foot down on the accelerator after a certain point it is not the pressure that increases the speed but the colour of the dashboard. And it is this colour that leads to the car going forward faster!

In fact, a focus on driving the speed higher by pressing down only on the accelerator may lead to a deceleration of the car because you will focus not on the experience but an element in isolation: 'the speed ticker' that only offers diminishing returns and a risk averse mindset.

So, to put it simply, because subjective experiences are qualitative, we need to focus not so much on the number as the meaning. After all, as we saw previously, we can have:

Consumers of Harrods score the shopping experience 8 out of 10.
Consumers of Sainsbury's score the shopping experience 8 out of 10.
Consumers of Aldi score the shopping experience 8 out of 10.

Same score different meaning, different return!

In setting KPIs then we need to be consistent with the qualitative nature of experience! Not reduce customer experience KPIs to a single number! After all such an approach may look simple and understandable but try answering the question why, and what should we do; that just takes us back to complexity: unless that is you don't care and are only interested in reporting data and ticking the box.

Now let's consider what this qualitative approach means in terms of outcomes such as growth and sales, because that's what we are trying to manage.

Management Implications

1. Subjective experiences emerge and modulate; they are not just about root-cause direct effects on behaviour.
2. Beware of response bias; this is used to fool you into a focus on efficiency.
3. Subjective experiences are fuzzy which means they behave differently and represent a more open system than objective data.
4. Subjective experiences are not entirely recallable.
5. Subjective experiences need context; sometimes 5 out of 10 is all you need!
6. Edge augmentation and figure–ground effects are an analogy for the impact of subjective experience, which means we should focus less on fixed effects and more on measuring and managing the flow of experience and look for signs of where it is heading.
7. Subjective experiences are intrinsic.

8. Subjective dashboards are different. We shouldn't be hidebound to follow the 'more of' philosophy. Frequently our most valuable course of action is to change the meaning of a number not drive it higher based on what is, rather than what could be (see Footnote 3).

Notes

1. I also note that there are other issues such as the validity of customers responding unnaturally to a survey and how representative this is of the population who aren't survey primed.
2. Being oblique in questioning is also important. Too constrained and you guide response, too open and responses become too abstract.
3. For instance, if you look at customer stories around a score such as 8 out of 10 on NPS you will find a vast difference in emotion words used and drives expressed about the score and around how customers perceive you.

6

The Subjective Data Line

With All This Fuzziness, What Does the Subjective Data Look Like? Is It Different From How Objective Data Relates to Growth?

The answer is yes, let me explain:

The more you put your foot down on the accelerator, the faster your car moves forward. It is a linear, objective, root-cause effect. However, customer experience data being subjective, behave differently; it is what's called curvilinear. Take attitudinal scores and their relationship to company growth, as shown in the example below (Fig. 6.1).

At the lower levels, 0–5 out of 10, if you increase the score, growth increases. Here, the fewer things go wrong the more likely your customers will stay. Negative things are also specific. 'I don't like the experience because of X.' This is the area I call loss aversion. And because this area behaves a little bit like objective data most of the focus of companies has been here; focusing on removing complaints, reducing churn, removing detractors or obsessing about hygiene and the provision of a perfect tech platform.[17]

98 Customer Experience Management Rebooted

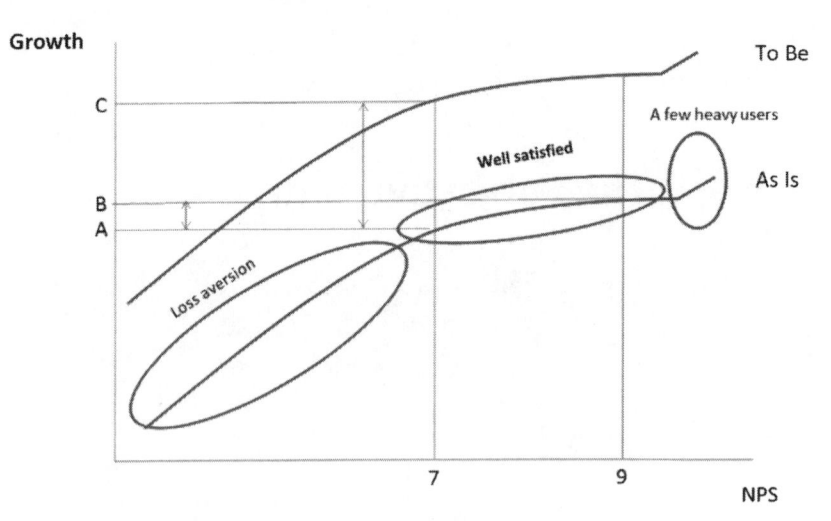

[Figure: Perception data curve with axes "Growth" (vertical) and "NPS" (horizontal, marked at 7 and 9). Curve shows regions labelled "Loss aversion", "Well satisfied", "A few heavy users", with "To Be" and "As Is" trajectories, and horizontal levels A, B, C.]

Note:- the curve is referenced in part to Professor Stan Maklan and Cranfield School of Management; with my take on it.

Fig. 6.1 Perception data curve (This example shows the general effect. I accept some goods such as luxury products may flatline at a higher score, but the point is the data are curvilinear and not about 'more of' a score all the way to 10 delivering higher growth. As we show even 10 out of 10 is not consistent in meaning!)

But just focusing here will only take you so far because only a small percentage of your customers are affected by loss aversion. (Let's not confuse this with ease because memorably easy is possible.)

Now let's go further along the line. At say 7 out of 10 the relationship FLATLINES. Why? Well, many experiences are just 7 out of 10 kinds of things. Like electricity and gas purchase: here even though you might invest to make the experience 'better' customers will still call it 7 out of 10, because that's what it intrinsically is.

Hygienic experiences act like this hence if you are investing in these things in your experience programmes beware: 'more of what is hygienic may mean nothing at all' and your resource allocation may be better spent elsewhere. Although you can always game the data to 'say it isn't so.'

Here we also find that many experiences that score 9 out of 10 do so through high use and familiarity not because the score drives higher growth. We can see this in the work I did with a large multinational (MNC). Here over 50% of the consumer audience scored 9s and 10s but at the same time a third were willing to state, when they thought about it, that they would never ever in part or in full recommend the MNC. In other words, the low churn and high use led them to tick a box that should have meant high recommendation when in fact it didn't at all.

Next time you score someone 9 out of 10 think about this effect. Most likely you gifted the score; in the moment it was OK but it didn't really mean you would actively recommend the company or even think it was any more than a nice satisfactory performance. And its link to behaviour is pretty slender too; the next day the score and event are forgotten because they are not meaningful to you.

Of course 9s and 10s may also score this just because they are luxury products.

Hence we see only a slight movement in growth with a move from 7 to 9 out of 10. This is shown by the move A–B, the relationship to growth flatlines. This is the area I call 'well satisfied'. Customers are 'well satisfied' and pushing for a higher score would be a waste of time.

Then there is the top end of the curve.

Here there is a small uptick at 10 out of 10, which relates to heavy users. These are a small subset, not a desired category, which would be impossible to reach anyway. Sure some of these people might get an 'I love Harley-Davidson' tattoo, but you would be fooling yourself if you tried to get everyone to behave like this! And it would cost you the earth.

In addition, there is a tendency for these people to tick 9s and 10s due not to any emotional pull of loyalty but due to the effects of high use. In effect, high use makes it more difficult for them to stop buying whereas those with lighter use who have a higher tendency to tick the lower range of the scale can more easily stop.

Finally, as we have seen, we also find a high percentage of 10 out of 10s not really meaning it; many would still never, ever recommend!

This is the area I call 'heavy users'.

So what to do?

Well, if it's not possible grow by increasing a score, move horizontally. Why not move vertically? Change what a 7, 8 or 9 out of 10 score means to a consumer. That way you would deliver more value to your business. This I call *resonance*, and is shown by the gains A–C made when we shift the line upwards: a move from the 'as is' line to the 'to be' line. In other words we add value to the meaning of a score not increase the score per se.[1]

And if we take that view, why not prevent changes in meaning that could lower the whole line by proactively responding to events before they cause damage (closing the loop between an alerted issue and an action). This is called *resilience*.

So resonance and resilience are important to customer experience as is a focus on the well-satisfied consumer; the tipping point between gains and no-gains.

In terms of loyalty this approach also means that we should not believe that more of say NPS causes a rise in loyalty. After all, if I am only prepared to help you if you give me 5p each time then that's a pretty poor way to build relationship! But this is exactly what NPS and other attitudinal scores are doing. We are only going to put skin in the game if a metric is moved!

Sure its true to say that NPS, CSAT and CES are not a problem – promoters are more valuable than detractors. But do not assume a linear root-cause link between more NPS and more behaviour such as spend; the truth is it's actually increases in use and tenure that increase NPS. Hence, any attempt to move the metric after a certain point will only focus the mind on granular hygienic things for no benefit when you should be looking to move the curve through understanding use better and innovating, which means keeping pace with change and being resonant and resilient.

So What is Resonance and Resilience?

Resonance

To be *resonant* is to talk about drives and memory assets, a similar concept to advertising although here we are talking about the experience. So it's not so much the case that you associate a resonant experience, such as chocolates on the pillow in a hotel, with a direct increase in revenue any more than you would an advert. It's more that resonance like an ad keeps the experience in mind, waiting for the moment when it is required.

Hence, advertising spends millions of dollars on ads, not to deliver an immediate return but maintain the brand's market presence: 'The sales of a brand are like the height at which an aeroplane flies. Advertising spend is like its engines: while the engines are running everything is fine, but when the engines stop, the descent eventually starts.' (Professor Byron Sharp, How Brands Grow)

Resonance means creating positive memories that motivate. Nudge is an example: how HMRC increased tax returns by 10% by sending brown envelopes not white; they felt more personal. Or how McDonald's refurbished their stores; Burger King now looks old fashioned. So with resonance and memory clues we are getting into design thinking. We are starting to create motivational clues. But we should not forget that finding experiences that do deliver direct revenue can be a characteristic of resonance, for example, how LUSH created experiences that opened up new markets. And we can up to a point measure the correlative effect: are we creating a sense of trust in a bank. But get too granular in your metrics and you focus on efficiency and lose a sense of how things emerge.

Resilience

Here companies are proactive in monitoring and being able to respond to how markets change. They go with the flow, become agile. Resilience means we need to be able 'to see' these changes otherwise we miss the developing threat or opportunity. So if you want to be an Experience brand you have to be open to creating resonance while monitoring the dynamic flux of resilience.

How Do You Manage For Resonance and Resilience?

In short you manage the curve not move the metric: at least after a certain point. This contrasts with firm efficiency approaches that focus on managing small changes in the metric (usually achieved by reducing detractors). Furthermore, by focusing on the curve, we are more open to market shifts. Efficiency brands by contrast will remain flat footed, a critical error in todays dynamic marketplace.

And the classic example of this is coffee houses.

Resonance:

Pre Starbucks, customers gave them 8 out of 10 and spent say $3.50 for a cup of coffee. After Starbucks customers gave them 8 out of 10 but this time spent say $5 for a cup of coffee. Same score, different meaning, more spend and more use! Coffee houses meant a little bit more qualitatively.

Resilience:

With the market changing, however, some coffee houses would stand still. Their clientele would initially still give them custom and 8 out of 10 scores, but now this is leading to lower spend and lower Share of Wallet (Fig. 6.2).

The flat-footed but efficient old-school coffee house has sat on its laurels and been unresponsive to market shifts. And I bet they made their biggest profits just before their market fell off a cliff as the new coffee houses expanded the market share!

Flat-footedness is also seen in how your employee culture also becomes myopic; your KPI in your silo this month is to move NPS +10.

Or worse, in my silo, I only care about getting my NPS score up +10 points and I couldn't care less about your target, which leads me to a focus on firm efficiency, even though the customer sees the journey and not your silo and even though +10 while looking good on paper really only means a tiny move in average recommendation for no overall effect.

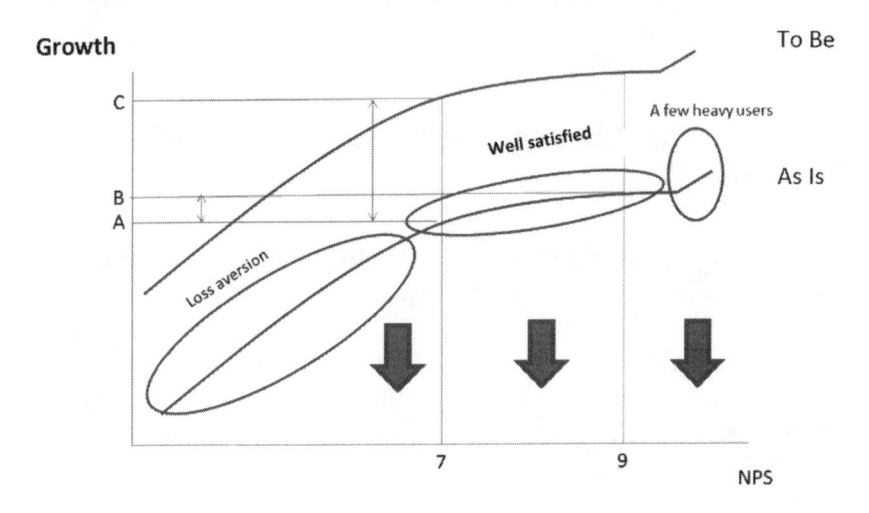

Fig. 6.2 Managing resilience

How Companies Game the Subjective Line

This line is probably the most important line in customer experience. It changes the notion:

- That linear rewards are always to be had from metric improvements (the 'more of' philosophy)
- That you should always just focus on what we do today even though you are not serving up an experience
- That waste is a bad thing, when it is part and parcel of adding value and taking the risk
- That you should not measure how markets shift dynamically but depend on a root-cause, closed system model

Yet companies tend to ignore this or game it by making subjective data behave as if they were objective.

Let me explain further:

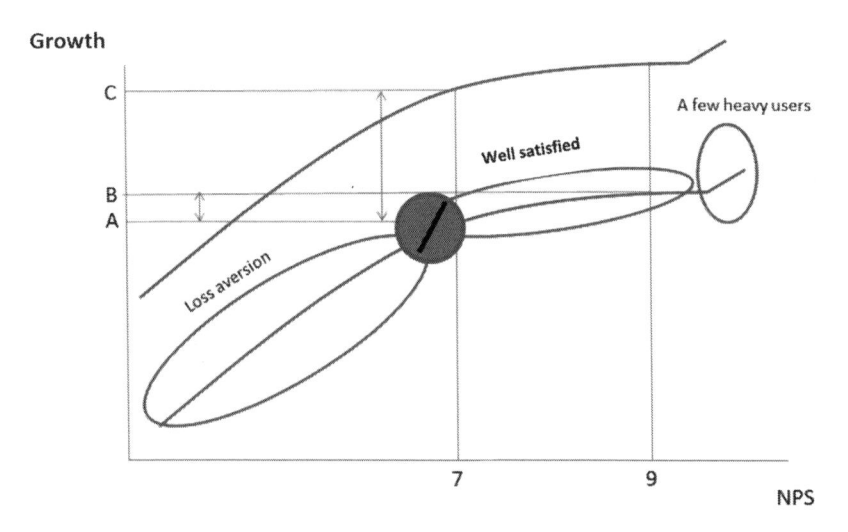

Fig. 6.3 The blue dot effect

The Blue Dot Effect

The blue dot represents an experience you as a manager are interested in. For instance, you might be targeted on increasing the NPS of floor cleanliness and prove its driving growth! Now imagine floor cleanliness is one question out of 20 on your overall brand NPS survey you send out to a sample of customers (Fig. 6.3).

What would happen under this bonused objective? You know that other effects like price will overwhelm such a small generally hygienic experience and you also know that customers don't notice such a granular thing, unless it goes wrong.

So what do you do? Now I have mentioned this in brief earlier but I think it is so important it is worth re-emphasising:

Firstly, you would measure floor cleanliness in isolation from any other things like price. This is done by sending out a survey asking customers for their NPS score on just this experience. Customers will see the survey and because they will always give you an answer will respond in one of two ways:

(A) They give you a low score because they have noticed how dirty the floor is. (5/10)
(B) They never really thought about it but since you ask will gift you a score based on brand. (8/10)

Then you use regression techniques that assume items like floor cleanliness are seen independently from other items in the experience. That is to say you mathematically remove the fact that customers don't see floor cleanliness separately from their brand score!

Then you don't mention the fact that the items only explain a small difference in brand NPS scores. (Say 5%, in other words other things matter and influence the score. Things we haven't included.)

Then you ignore the fact that scores are frequently forgotten and meaningless; and no you cannot use some imagined depreciation approach, that assumes a score degrades through time in some algorithmic way.

Then you use the fact that customers gift scores (say 8 out of 10 or more) and that we do get some negative effects (say at 5 out of 10 or below) to draw a line between the two points and extrapolate all the way to 10, ignoring the curvilinear nature of subjective experience.

I show this with the blue dot. In reality it is static, but the model shows it as a linear effect.

And if you do all this, congratulations! You have put the assumptions of an objective statistical model before the data and kept your job and bonus; after all it must be true, it's objective maths.[2] And we haven't even got to the fact that you could replace cleanliness with a hundred other items in isolation and get the same level of explanation (5%). In this way you explain 5× what is there: something isn't right! Or the fact that things such as NPS inflate effects: a +10 may not mean as much as you think it does. Or how you have failed to capture the changing dynamic of the market, because you have pre-canned the questionnaire and ignored modulating influences. Or that customers have a non-conscious mind. Or that rather than evaluate, they signify what is important to them descriptively. Or that they don't just use the past to predict the future as much as construct various If-Then scenarios for their spend, forward predicting what might happen and deciding with emotion - using past information but also other salient scenarios.

Management Implications

1. The subjective data line is different than the objective data line: demonstrating loss aversion, well satisfied and heavy users.
2. Resonance and resilience are crucial to our understanding of experience. They focus our metrics and KPIs on managing the flow of an open system where we need to look for dispositions.
3. Manage the curve not move the metric! At least after a certain point. Otherwise, you will obsess on granular changes for no effect and miss the bigger picture; how the market itself might be shifting.
4. Companies game the data using response bias and assumptions they fail to declare.

Notes

1. You should seek more of something if you are a poor performer. If you score a poor 5 out of 10 then you should try to raise your score. However, if you are close to a plateau, say you score 8 out of 10, then expect diminishing returns from a 'more of' approach! However, there is another option. Rather than try to achieve a higher figure, seek to change the meaning of that figure. We could make a score more resilient. So, network reliability may score 8 out of 10 on NPS today because the brand is well known and it might score 8 out of 10 tomorrow but be more associated with network quality! And hence drive more value. Likewise, rather than get shoppers to go to a grocery store and score 9 out of 10, accept that they score 8 out of 10 but make the meaning of 8 out of 10 deliver more value. For instance, 'I' go to the store and the product I wanted is on the shelf, but now it has accessories. It's not that I score more; it's just the same score is now more resilient because you offer more of the product I want. Or 'I' go to the store and I see a cooking display. I still score the store 8 out of 10 but there is more resilience in my association of the store with cooking products.
2. The irony here is that NPS does not measure dissatisfaction and a bipolar CSAT scale means consumers answer towards the neutral because a call centre has aspects of satisfaction and dissatisfaction, or answers tend towards the positive because you ask me about satisfaction! And I gift you an answer.

7

Customer Experience Is Complex

When you go to buy a cup of coffee you don't look at all the functional aspects of purchase before you make a decision! You don't say: let me evaluate the quality of the cup, the temperature of the coffee, how quickly it was served to me and so on and so forth all on a Likert 0–10 scale which I then convert in my head into a regression model to come up with a result.

No, managing subjective data is not like building a car; it is not an engineering project.

Instead, it seems as humans we do something else. We respond with feeling, thought, a conditioned response: 'I'd love a chai latte,' 'I like this store,' 'I fancy a different variety today,' 'Hey that new coffee shop looks different; it has an odd name "Caribou Coffee"; let's try it.'

So although I accept we can and should up to a point correlate some aspects of the experience with NPS and spend, I really do believe that to treat customers as cost-benefit calculators measuring against 'as is' only expectations is constraining.

The implications of taking too much of a 'calculative' view of customers is that you will, in my opinion, reduce your capability to understand the qualitative, sensory, nonconscious, phenomenological experience. So although with the calculative method, you will understand that, 'trust is important,' 'reducing the price of the bill,' 'the reliability of network connection,' you will also fail to get down to a deeper level of understanding of why we do what we do.

Or what else is or could be missing. Think of the 'compare the meerkat' advertising campaign and its effect on growth. Before the campaign, 'interest in adverts' would surely not have been a correlate to spend!

It seems therefore that we do not so much have a ruler in our heads by which we process information against prevalent conscious and nonconscious needs as receive new information from which we construct a sense of the experience alongside our prior expectations and drives

So, on holiday in Hunstanton, we go to a new beach; our expectations of need achievement being set by our past knowledge of beaches.

When we arrive, however, we don't measure our experience using some ruler. Instead we construct a sense of the experience by reference to our intuitive need expectations and more importantly from new information we receive in the moment, because this beach offers a qualitatively different experience. Not forgetting of course, the impact our own personality, mood and so forth has on this process.

Hence, it is this phenomenological reaction to the experience as it comes to us that makes our understanding of CX different from traditional statistical-based approaches.

So rather than say this beach is 8 out of 10 based on prior need expectations we say:

> 'When we went to the beach we liked the walk through the pine forest. On arrival it was very different than expected. A beach with dunes around a lake you could wade through. My kids loved it. We then walked for 5 minutes through this area to a beautiful beach, hidden away and remote with no encumbrances from shops or roads.'

But in spite of all this many companies remain wedded to a statistical root-cause approach. You only have to look at the huge number of voice of the customer solutions on the market and the wide array of promises of return to realise that! I mean if customer experience and NPS were like measures of web browsing speed, we would have a set process, a set methodology that would be perfectly predictable.

Now I think in part companies realise this. But the reaction has been a disaster. It's fluff, make the data hard, 'subjective data' cannot be managed so let's ignore them. For me, if you take that view, you are logically inconsistent. So you are interested in the experience the customer has, but you want to ignore the experience the customer has, so why are you interested in the experience the customer has?

Of course, that doesn't mean we shouldn't review objective analytics to see 'what customers do' and impute their reaction. After all if someone 'looks' confused on a website, it's reasonable to assume they are; if someone downloads music by Lady Gaga, it's reasonable to assume through recommendation engines what they might want to receive next. But even here, we need to add in subjective data otherwise we end up myopic (we don't see the intangible drives) and miss the elephant in the room: the 'why?' And the 'what could be?' Maybe I hate receiving these recommendations (or don't care); maybe you make the website a great user experience but I 'want' something else and the look and feel is rubbish.

This is why I believe a paradigm shift is required. We need to move away from engineering and financial metaphors in our research. Customer experience is complex, not complicated, to use the language from the Cognitive Edge, Cynefin framework.

We need to consider subjective data and change how we measure and interpret the results. Let me explain.

The NPS Question

Question: 'As a manager you have been tasked with improving NPS; how do you approach this question?'

Faced with this problem the traditional manager would think like an engineer. This is unsurprising because accounting, command and control, TQM and other business approaches depend on this style of thinking. This is called the complicated approach.

> Complicated systems – systems, in which there is a cause–effect relationship, but it is difficult to detect it. Finding a solution to problems from this

domain often requires expert knowledge, a lot of experience and complex analysis. Apart from that these are usually static systems or systems that are not really vulnerable to changes (if there is a change, you can easily predict and analyze it). Examples include a watch, a car or a house – they are static, but to create or repair them we need some expert.

http://msieraczkiewicz.blogspot.co.uk/2014/02/simple-complicated-complex-and-chaotic.html

In complicated thinking there is a direct analogue between input and output so in an everyday example: of driving a car (Fig. 7.1):

Here, effects are predictable. The car will move forward when I press my foot down on the accelerator.

With NPS improvement as an objective the manager would therefore think like this with the one proviso that because human interpretation is involved the model would look as follows:

NPS equivalent model (Fig. 7.2)

However, we don't see this as a big deal because all we need to do is use surveys to ask the key questions pertinent to the experience the customer has, then we can predict any rise in NPS and hence a rise in spend.

Here is a call centre example:

As a manager we bring in a survey used by many call centres to understand what the drivers and destroyers of NPS are; this short survey might look something like this:[1]

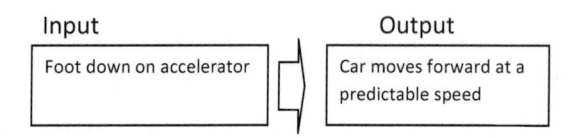

Input	Output
Foot down on accelerator	Car moves forward at a predictable speed

Fig. 7.1 Complicated effects

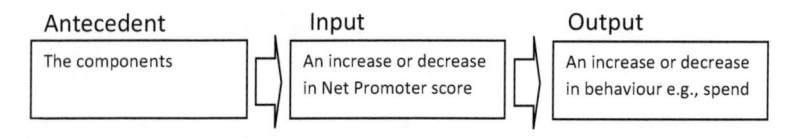

Antecedent	Input	Output
The components	An increase or decrease in Net Promoter score	An increase or decrease in behaviour e.g., spend

Fig. 7.2 Complicated NPS

Please provide your recommendation score for:

1. Empathy of call centre rep
2. Length of wait time to speak to a rep
3. Friendliness of rep
4. Getting what I wanted
5. Ease of doing business
6. Overall NPS score for the experience

Then we have a supplementary question... Please explain why you scored that way.

We then use weekly correlations to inform our reps on what they need to improve. By using statisticians who understand our business we further analyse those consumers who score 9 and 10 out of 10 to understand what drives or destroys best performance.

In this way we obtain predictions; that is, if we increase empathy by 1 point we will get an increase in NPS of 10%. In fact we can go further than this; because we have expenditure data on the top promoters we can quantify this link to predict a 1% rise in revenue.

Of course we need to turn this finding into practical action, so we decide to look at empathy a bit deeper. Hence, we look at the verbatims from the question, 'Please explain why you scored that way', to find the reps with the highest average empathy scores.

By then focusing on rolling out empathy training so reps achieve the same performance or by getting in new staff who have equivalent levels of empathy we can now practically raise NPS, raise spend and understand how much it will cost.

In the same way of course we find 'negative moments' in the experience. For instance, looking at the data, it is clear that those customers who speak to poorer performers have a poorer experience. They are consistent underperformers on empathy. So we decide to let them go.

By managing experience in this way, like a machine, we can drive NPS improvements and grow the company. The manager at the quarterly meeting shows charts going upwards, gets a bonus and is in line for a promotion.

The company now decides to roll out the programme to all their call centres under the banner 'Driving NPS up!' Call centre agents are informed that their bonuses and chances of promotion are determined by their NPS scores. But more than that, this statistical approach will now be used in other parts of the business such as in-store.

And to be fair this feels partly correct. I am not entirely anti it, especially when it aligns a disparate matrixed company around a single goal and helps to shape up the account management structure.

For instance, you could envisage a call centre with a bland overall standard but some stellar performers who represent best practice. You could imagine replicating these people and seeing your performance on NPS move up. You could also see the benefit of taking out bad call reps who consumers hate to talk to and focusing on generating promoters of your business. You could see how building in a focus on customer survey data may encourage a customer-focused culture and reduce the impact of the cost-cutting CFO.

But hold on a minute. Be careful what you wish for; perhaps this is getting a bit out of hand?

Certainly some parts of the experience the customer has will operate in a predictable manner. For instance, if 'complicated' is defined by the ability to define a root cause, customers are up to a point going to be clear about what causes any lowering of their score; although let's be honest: some customers you want to get rid of!

But a call centre environment as with any part of the business is not just defined by managing what goes wrong! After all most of the time we are entirely satisfied with the experience.

So although the programme may be a good one, it remains constraining.

Indeed perhaps we find that the way we execute empathy has perverse results: scripting and a culture of 'no mistakes' reduces creativity and flexibility. Maybe by focusing on the number, gaming results – whether deliberately or inadvertently – we become blind to understanding what the customer actually wants. Maybe we recruit a team of reps from another industry where customer service is more important, under the belief that they will be empathetic. Only to find that the environment is so different that results worsen even if they are 'more pleasant'.

In this way, we may identify an action to perform based on historical data, but to execute a change means adding in something new and qualitatively different, with potentially unexpected and unpredictable results.

Also, perhaps empathy is important, but it's good enough already and in fact customers are more responsive to other aspects of the experience.

So we may find that what is more important is not moving a static metric but how we react to change.

I mean markets move, situations are always slightly different from one another, technology changes, brand perception alters, advertising offers come and go, we miss out on unexpressed nonconscious or even conscious emotions in our root-cause metrics. Processes go awry, that great rep on the phones suddenly goes through a divorce and no longer comes across well on the phone, a change of boss means we have a control freak in charge and now we are all unhappy, we react to change badly because we are all worried about our NPS scores, or that abusive customer is kept 'sweet' even though before NPS we would have just got rid of him.

And that the fixed survey is not a perfect representation of the experience anyway.

But still you run your driver analytics approach, assuming everything is perfectly represented in a static model and we can run everything like an engineering project.

To go back to our subjective data line, we seem to be managing the metric not the curve!

So to me the failure of customer experience stems from the fact that businesses have been using the wrong measurement system! The wrong equation! Why? Because customer experience is not so much complicated as complex! It is not so much fixed as constantly moving! It's not a number so much as a 'meaning'.

To quote Dave Snowden: 'Any supposedly objective assessment that tries to identify drivers runs the risk of false reassurance for the organisation, encouraging complacency or misdirected intervention.'

But what does that mean?

Complex Thinking

Complex systems are ones in which there are no clear cause-effect relationships, because they change with time. We can detect them through experiments and investigations into the current state of affairs. Even expert knowledge does not allow us to arrive at a solution, but we can use it to set the direction of investigations. Systems under this category are live, organic and changing. Where there are people, there are usually complex systems. Examples of such systems include a stock market, a brain, the immune system, societies. Acting in the space causes the space to change, and cause and effect can only be understood in retrospect.

Here's another commentary:

Complex adaptive systems show behaviour that constrains agents (e.g. customers, perceptions) but is itself changed by those agents' interactions. So customers' perceptions are constrained by company systems but those perceptions themselves change the company's approach. This, coupled with the many other interactions and feedback loops, means that seeing customer perception as anything other than a context-sensitive complex adaptive system is a fundamental error.
http://msieraczkiewicz.blogspot.co.uk/2014/02/simple-complicated-complex-and-chaotic.html

It's a bit of a mouthful which is probably why this view of customer experience has stayed in the shadows. So I give some admittedly simplified examples and explain some of the principles that make it different from a survey and better for managing that subjective data line.

Complexity Means Understanding the Flow

Customer experience being subjective means there is a flow of experience that managers need to keep watch on. They need to understand how things are moving, catching bad things early and amplifying the good, which of course costs less than waiting till something damaging 'has occurred' or missing out on 'something good' that your competitor has picked up on.

Here is an example:

I regularly go on holiday to Broadstairs. Three years ago it was easy to get a place on the beach and eat your ice creams and sandwiches. However, last year it was noticeable how, although the beach was still full, it was becoming increasingly difficult to enjoy the experience. This was because the seagulls were becoming aggressive, stealing food and generally being a nuisance. This is an experience that is gradually increasing the 'dissatisfaction comments' and although only affecting a few decisions now, will grow to affect the number of visitors.

In other words, a growing bad event is not visible in aggregate, but is visible by looking at individual incidents. Hence, by seeing how the system (the beach holiday) is disposed you can proactively manage bad experiences and amplify good ones. Not wait for something to become a destroyer of value before you act.

And we can also see this in how after the event we talk about it and reflect on it: all changing variables.

For negative experiences this is rather like checking an aircraft wing for hairline fractures and proactively prevent them growing into something more serious.

This is why complexity metrics set parameters; they are designed to flex and alert us to a change so we can adapt. In such conditions parameter metrics are better than targeted rises!

The interesting thing about this is the concept of 'alert' also fits with how emotion works! It tells us about events in our environment that are salient to our well-being and we need to respond to 'in the moment'.

Complexity Means Understanding Resilience

Resilience management is best expressed by my family visit to Charles Clinkard's Shoes in Milton Keynes shopping centre. Here if we were to run a survey, it would be clear. Our primary drive for going to Charles Clinkard's would be 'We want staff to know what they are talking about.' Any verbatims would say:

> We scored NPS 9 out of 10 highly because they are the only shoe shop that has a good variety in store and we get such excellent service from employees,

who know what they are talking about. They really take care with their shoe fitting; for instance shoes are always double-checked by a second member of staff. Although it is a bit crowded sometimes.

Any reduction in 'staff who know what they are talking about' clearly would reduce the store's value to us. But at the moment our score on NPS is 9 out of 10.

Now imagine the next time we go to the store we had a poor staff experience. What would happen? Well, perhaps we might still score 9 out of 10 and say it was 'one of those things,' but our confidence would have gone down a little and next time we would be more sensitive to changes.

No change in score, but it is becoming less resilient.[2] That curved line is starting to wobble. If it happens again we might start to say, 'You know it's a long way to go; there is another shop in another town which is easier to get to,' and so on and so forth.

Complexity Means Understanding the Employee System

But it doesn't end there. What about the employees?

For instance, we could take the Charles Clinkard experience and see it from their point of view. Now the bad experience was due to:

My kids were playing up and I was in a bad mood (changeable).
The employee just happened to be a transfer from another store for the day and didn't know what to do (changeable).
The employee had a bad day, that's all (changeable).
The shoes were not in stock and the employee was doing her best to find them, hence the long wait for the wrong shoes (changeable).
The employee was rude, because I was rude (changeable).

In training, the business may seek to take the words 'know what they are talking about' but knowledge from a book is one thing, how it's put across, body language, tone of voice and the ability to flex rules to the situation are another entirely.

Complexity Means Building Memory Assets Being Resonant

And then there is 'resonance'. Here is an example of how complexity comes into play here.

Car repairs are always a distress purchase; the sense that these guys are going to rip you off is always there. Hence, my Vauxhall garage experience of paying £1,000 for car repairs when I expect £200 led to a lot of distrust and a promise never to use them again or buy a car from Vauxhall for that matter!

You could almost smell the sales targets. The environment spoke of lack of care.

Contrast this with the experience I received from another garage in my local area, themed to express 'old fashioned values' of care and the £1,300 repair bill felt 'just one of those things.' The fact that they were open and transparent and let me see the repairs also helped.

Complexity here means resonant clues that indirectly affect how we feel about other things that happen.

So, higher repair bill, same car, and same low level of knowledge about the repair but a different response entirely.

However, take care! Resonance also builds emotional sensitivity. A few months later I saw an advert in the newspaper that went against them; now the fact that the garage was empty when I went means something different. My cognitive recollection changes to other aspects I previously hadn't noted or thought important.

Finally Complexity Means Trial and Test

Remember the concept of the modulator! When things influence each other, the only way we can see if we have moved the dial in terms of how customers think, feel and behave is to trial and test.

To give an example, in a habitual experience like shopping it might be possible to say customers are influenced by price and variety of goods on sale. Now if we decide to increase the variety of bread and dairy products on sale we really will not know if this increases NPS and spend unless we trial it. You can't predict whether sales tomorrow will work unless you do it.

So a major grocery store during the recession had a hunch that consumers would like to get an all-in-one box so they could make their own meals cheaply. On the face of it, with consumers so conscious of price it sounded great. In reality, testing it on the shelves, it failed. Consumers didn't like the look of the goods, felt patronised, felt it didn't fit with the store and so on and so forth.

Likewise, if you run a churn analytics programme on perception data, you may come up with five great campaign ideas, but you don't really know whether you will decrease churn. You can surmise of course, but remember markets are changing, influences on churn alter, and a campaign's success depends on what you do and how you execute it: which is not necessarily about one possible solution to one problem. So once again the trial will be the only way to know.

And trialling is not unusual…

Tesco do experiments, the pharma industry is well-known for randomised controlled trials, and in the show *The Fixer*, expert hunches are always tested out.

And experts are important, because they are the ones to come up with the ideas (and I don't necessarily mean consultants here, I mean consumers, suppliers and employees).

So you should not depend on an 'as is' mathematical model. Experience design is as much about trialling and testing hunches as it is about monitoring the flow of customer stories.

By contrast if we rush to ROI we will turn our intrinsic offer into something less valuable as we seek to manage the metric not the curve.

An example of this is from the Leela Hotel in India. Here they started to cut corners to save money. When rooms were left empty during the day, the air conditioning was turned off and some of the beautiful displays of fruit were taken down to save time. But these small actions were intrinsic to value.

Of course there is no ROI for each of these actions individually but by undertaking to remove them, the hotel was heading down the line of reducing the value of the hotel. Critically, this wasn't about managing everything, the weak signals of failure (fruit displays and air conditioning) were identified and quantifed by customers.

And we shouldn't forget the cultural feedback loop in this hotel example. These cost-cutting activities are the thin end of the wedge; if successful they can become bolder leading to a definite degradation of 'experience' and employee investment in the experience.

Management Implications

1. Subjective data are not like building a car; they are not an engineering project.
2. We respond with feeling, thought, a conditioned response.
3. Taking too much of a 'calculative' view of customers is that you will reduce your capability to understand the qualitative, descriptive, sensory, nonconscious, phenomenological experience.
4. NPS is only a partial customer experience metric because it is designed for root-cause, complicated approaches.
5. Subjectivity is more complex than complicated which means it is less amenable to root-cause analytics and more amenable to managing the flow.
6. Be aware of dispositions: amplify the positive, reduce the negative.
7. Do experiments: trial and test is the only way to know if something will deliver ROI in a complex system.
8. Any supposedly objective assessment that tries to identify drivers runs the risk of false reassurance for the organisation, encouraging complacency or misdirected intervention.
9. Remember the employee and other stakeholders' experience.

Notes

1. The irony here is that NPS does not measure dissatisfaction and a bipolar CSAT scale means consumers answer towards the neutral because a call centre has aspects of satisfaction and dissatisfaction, or answers tend towards the positive because you ask me about satisfaction! And I gift you an answer.
2. You should seek more of something if you are a poor performer. If you score a poor 5 out of 10 then you should try to raise your score. However, if you are close to a plateau, say you score 8 out of 10, then expect diminishing returns from a 'more of' approach! However, there is another

option. Rather than try to achieve a higher figure, seek to change the meaning of that figure. We could make a score more resilient. So, network reliability may score 8 out of 10 on NPS today because the brand is well known and it might score 8 out of 10 tomorrow but be more associated with network quality! And hence drive more value. Likewise, rather than get shoppers to go to a grocery store and score 9 out of 10, accept that they score 8 out of 10 but make the meaning of 8 out of 10 deliver more value. For instance, 'I' go to the store and the product I wanted is on the shelf, but now it has accessories. It's not that I score more; it's just the same score is now more resilient because you offer more of the product I want. Or 'I' go to the store and I see a cooking display. I still score the store 8 out of 10 but there is more resilience in my association of the store with cooking products.

Part III

Customer Experience Research

In this section, we now look at what we can do to build best practice 'customer experience research'. This underpins how we can start to create and design an Experience brand, because if we have the right measurement system we can move beyond efficiency.

One key focus here is how experience metrics look to understand the flow of experience not bind it to an efficiency engineering type model, and in that way influence the direction of travel, being sensitive to how things are changing and invest in modulator effects (small important changes at the edge).

8

Traditional Surveys Are Efficiency Surveys

Traditional surveys play a crucial role as thermometer checks across key experiences. So keep them! They ensure quality thresholds are maintained.

But…

Traditional surveys are NOT customer experience surveys.

They are not the same as the experience the customer has.

Hence, there is a need to complement them with something more valid to 'the experience the customer has' and less prone to a number of biases that arise when you ill consider the unit of measurement – the human mind.

Here are a few of the problems with traditional survey techniques that to my mind make them efficiency surveys, working best to identify the things that go wrong, not the value of what goes or could go right.

They Assume a Complicated, Static, Engineering Logic

Traditional surveys serve up a series of questions that define what the experience is! The problem is this assumes that the experience the customer has is an inflexible closed system defined by mathematics (the scale and statistical calculations) and that other impacts do not matter. In effect the survey sets a static frame based on what the firm wants to hear (Fig. 8.1).

As an example, on a BA flight from Delhi to Heathrow, I decided to spend 20 minutes running through their standard customer experience survey. This was scaled from 0 to 10 and asked a predefined set of 20 questions. The problem is, these questions were not at all representative of my internal view on 'what my customer experience' was. Now of course if you give me a set of questions, I will quite happily give you a

Traditional View   Customer Experience View

Closed system defined   Open system not
by 20 survey questions   defined by a static
                                set of questions

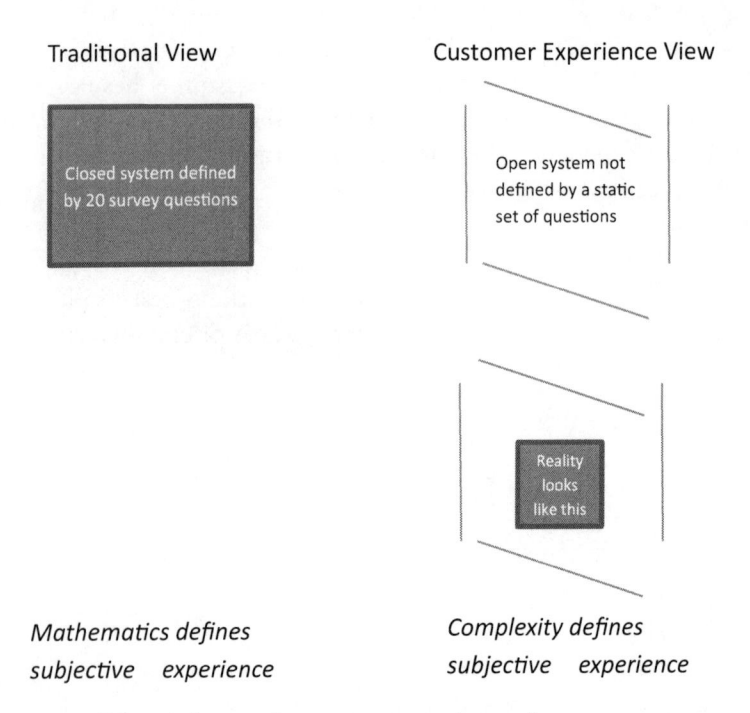

                                Reality
                                looks
                                like this

Mathematics defines   *Complexity defines*
subjective experience   *subjective experience*

Fig. 8.1 Traditional view and customer experience view

scaled answer. Worse than that, my answers will also reflect 'what feels the right thing to say'. Hence you get a false positive result.

That doesn't mean to say the survey doesn't ask important things about say seat comfort. But sometimes, frequently even, it is enough, not differentiating and other indirect, nonconscious and changing influences matter such as the well-stocked or otherwise music library, or how staff interact with me. Let alone social desirability bias. And let alone the fact you can have two diverse responses to the same 'instance'; that is, seat comfort is both positive (nice cushion) and negative (shame about the leg-room).

But instead of accounting for this surveys often miss these impacts.

Even worse, surveys operate constraining statistical processing regimes: assuming customers sum all experiences, that any given score is fixed and directly representative of what the customer thinks and feels and that customers act mathematically, having some calculator in their heads by which they measure everything against expectations and then mentally process data in the same way as a regression model! As one CEO said to me, these things assume life is a static picture rather than a film.

So consider this when we use scaled surveys to measure experience:

Could you exchange all the questions on your 20-minute survey with others and still get the same level of explanation?

Could you find that some statistical model is in fact poorly predictive beyond efficiency gains? Except of course in retrospect, where you can game the data!

Could you find that measuring everything just delivers a deeply circular model based on measurement error and customers who gift responses?

Could you find that focusing on 'as is' we are missing out the creative potential of experience because what 'could be' cannot be predicted very well? Even though we make decisions based on running If-Then forward thinking and feeling scenarios.

Could you find that customers have unmet drives? Hence, any survey or driver analysis of what 'is' would not pick this up: a focus on 'as is' would be a focus on under-performance.[1]

Could you find that we miss out the experience the customer has, because customers do not calculate and sum everything?

Could a score on a scale be unstable and subject to change; it's fleeting and forgotten (although seems permanent on a datamatrix) as well as not mean what you think it means – after all it is just a tick in the box?

Could asking questions abstractly outside a decision moment, just lead to abstract answers?[2, 3]

So sitting in your arm chair, how important is reliability to your brand NPS score? Maybe it's not in the relationship that's important when we measure, but in the decision moment[4] (see Footnote 2).

Could your transactional survey be deeply myopic, missing out on how customers don't notice the event and how they are influenced by other aspects of the experience, which are frequently nonconscious and emotionally fleeting? Such as, I didn't notice the IVR system but I hate your bills. Or I am deeply influenced in my response by prior brand beliefs even before I interacted with you!

And my doubt about scaled surveys also extends to text algorithms. Perhaps these too fail to measure experience effectively?

For instance, I might write about an Apple experience and say bad things, but 'I really like the place.' So although I appreciate that in mass data feeds from social media it is difficult to get the customer to 'rate and state' what they mean from the surface layer of a verbatim, in surveys, just get the customer to do this! Then you are 100% guaranteed to know what they mean.

Another example of this is the word 'bad'. Take 100 customer comments with the word 'bad' in each. It is not possible from a text algorithm to know the intensity with which each customer said this, if they even meant it, or if mixed with other contexts to what extent this is the salient feature or is impactful on value. But you can if you ask them.

Of course, don't let me criticise too much though. Watson analytics and other such approaches work where text is static and literally means what it says on the page rather than being hidden or requiring a personal interpretation. This also works when dealing with data that do not operate in an open system. So medical documents, unlike people, do not act as separate agents that interact with each other and change behaviour as they interact! Hence they are ideal for AI.

Also, working on a 60:40 rule, I don't deny we get something from text and sentiment analytics.

And what of emotion, conditioned response, unmet drives and nonconscious peripheral effects all made 'in the decision' moment. Do you think traditional approaches capture these well?

I mean surveys are often delivered weeks after an experience. By that time, many of the in situ effects that influenced you will be lost or rationalised out; for instance emotions are fleeting, the impact of how the salesperson spoke to you in the store, your mood on the day because the sun was shining, all these indirect and direct influences critical to your in-the-moment decisioning are lost.

Surveys and verbatims also fail to measure deeper drives. For instance, I may say I'll go to Starbucks for a coffee but in reality I want to sit down and talk to friends (the Third Place); it is less about the coffee. A standard list of items in a survey will not only fail to uncover these deeper drives but will also make us blind to them, by giving us a false sense of knowledge.

My fear of course is that confusing complicated approaches for complex will lead to a belief that customer experience is somehow hyperrational when in fact we act on fragmented knowledge, hunches, intuition, emotion as well as cognition, fed by knowledge of ourselves and memorable only if it is meaningful.

For me at least, we need better customer experience research! Which I now outline based on complexity science principles.

Management Implications

1. Traditional surveys play a crucial role as thermometer checks across key experiences.
2. Traditional surveys are NOT customer experience surveys; they fail to pick up on its phenomenological nature.
3. Surveys operate constraining statistical processing regimes: assuming customers sum all experiences.
4. If you hadn't asked the survey, would it still be an experience?
5. Emotion, conditioned response, unmet drives and nonconscious peripheral effects all made 'in the decision' moment are missed by surveys.

Notes

1. Indeed, one of the claimed best examples of customer experience comes from the Royal Bank of Scotland. Here they brought in an engineer to manage the programme who focused wholly on cost reduction, Lean and Six Sigma. The objective became not 'What can we do for the customer to increase the subjective asset, customer experience,' but 'How can we benefit ourselves, cutting out processes that seem superfluous.'

    The case study was a great success but was it from a customer experience point of view? Yes it reduced cost but my question is: Did it deliver an appreciable return on subjective experience? Furthermore, who is to say that some of these invisible cuts might well have been important components of a customer's 'experience,' or set the platform for more experience creation? I would contest that without a valid multimethod measure of experience no one could tell. But its use of activity-based costing is a significant addition to the CEM literature.
2. Stories must be picked up relevant to the situation. In other words not just an abstract 'Tell us your story' question, but 'Tell us your story related to purchasing a phone, recommending a friend and so forth.'
3. Being oblique in questioning is also important. Too constrained and you guide response, too open and responses become too abstract.
4. I also note that there are other issues such as the validity of customers responding unnaturally to a survey and how representative this is of the population who aren't survey primed.

9

Best Practice CX Research Methods

The problem with customer experience research is that the unit of measurement is the human brain. This means we cannot measure things in the same way as we would for something objective. For instance, an event such as web browsing speed can be measured by a network probe, the output of which will be data on a chart which exactly matches its analogue: web browsing speed.

By contrast, neuronal activity in the brain at the moment a decision is made cannot be measured except by self-report, which means the method of measurement affects the result! And is usually not an analogue for what it is measuring. As an example, measuring NPS means what exactly? That's not the same as web browsing speed! NPS is an abstract quantity with multiple changing influences.

Hence this system is rather like quantum physics!

You can never know the speed and position of a particle and if you seek to measure these things with hard objective physics, they will collapse into a new state. In the same way, you can never know exactly how customers think and feel and how this determines behaviour; if you seek to measure these things the data will collapse under response bias.

For instance, imagine you sell software. You now want to 'prove' its impact on NPS. So what do you do? Well you run a survey asking customers about their score on the few items you are interested in. A customer then gives you a score of 8 out of 10; but that's just because I asked you the question. For the vast majority of your customer base, it may not be salient, not important (unless it goes wrong) and not noticed. Priming for some granular functional item here misses the bigger picture: that perhaps other things are important, or perhaps I don't think about you very much at all.

So we end up with two realities: the reality of what really goes on which we have misconstrued and the reality of the software firm which believes it is oh so very important.

So what to do?

Well I think like quantum physics you should undertake a series of 'weak' measurements that probe the system but don't collapse it. And these weak or multimethod measurements are 'customer narratives', 'immersion', 'depths' and expert judgment. But ultimately, the controlled experiment.

Researchers therefore need to stop thinking like engineers and mathematicians; customers don't!

Customer Experience Research 1: Use Narrative Metrics

So what is the language of customer experience? What should we be measuring?

Part of the answer I believe should come from narrative- (or story-) based research because narrative, based on the thinking of Keith Oatley and Dave Snowden/Cognitive Edge, is how humans remember and engage with experiences. Narratives also bring together our emotional and sensory impressions, both powerful influences on how we make decisions.

To illustrate the importance of narrative, I refer to a passage from Henning Mankell and his observations on human behaviour from his time in Mozambique:

It struck me as I listened to those two men that a truer nomination (name) for our species than Homo sapiens might be Homo narrans, the storytelling person. What differentiates us from animals is the fact that we can listen to other people's dreams, fears, joys, sorrows, desires and defeats – and they in turn can listen to ours.

In CX terms this for me translates as follows.

No customer walks out of a store saying that was a great 8.5 out of 10 experience; what they do is tell or recall a narrative, such as 'It was really difficult to find a parking space.'

If you wanted to go to a restaurant you might say to yourself, 'If I go there, then I will get a great pizza and the kids will love it.' And allied to that will be a remembered impression of the atmosphere and what it felt like.

If I went to a mobile phone store, my need might be to get a cheap price but my narrative will give deeper and richer statements of what drives me, such as my concern or otherwise about the price, how the shop treats people interested in a low price phone, whether I will be ripped off, the crowdedness of the store and so forth.

Narratives then are more valid to how subjectivity works and inform us of how things are changing, picking up small clues and dispositions in the experience on offer, important in a fast-moving increasingly digital environment.

Narrative also has the advantage of shareability. Unlike histograms and pie charts, the effect of a good narrative on employee understanding is stronger. For instance, this is how Amex spreads the message on what it is like to be a customer.

A feature of narratives is also, according to Dave Snowden, that they are 'fragmented and fractal.'

For myself, I think of it this way. Imagine all the customers in the United Kingdom for British Airways. Now, if they were asked in let's say a recommendation situation[19] to come up with what they would remember, those millions of customers are likely to say similar but slightly different things. Of course if we asked a survey we could standardise things but I would bet that those standard items would only partially represent the unprimed memories of BA that affect how customers think and feel.

So what we also do is, using Cognitive Edge principles, get customers themselves not a text algorithm to quantify their narrative (not a pre-canned list of items) and set up a KPI to measure and manage customer experience: that is, 'More stories like this, fewer stories like that' (Source: Cognitive Edge).

After this, we then interrogate the 'hotspot' narratives picked up by the statistics.

And sometimes we find that those hotspot narratives may reflect something else other than what we were measuring even in a trans-actional event: such as how customer service or the price plan affects impressions of mobile signal reliability. Essentially, a bad plan makes us more sensitive to any reliability issues we would normally forget or we have a bad plan and that means we also see reliability as bad; it justifies our position.

Indeed this is what I found when I ran probably the first narrative NPS study using complexity methods.

Or how some hotspot narratives may only be mentioned by a few people but seem very important when we chunk the numbers, because like the Broadstairs example, they indicate where the flow is going or some hidden meanings that should be amplified or alerted, or how things are being used in a different way: an exaptation.

Here is an example of this:

Sometimes, I go to the Black Horse restaurant in Woburn. One of the key moments is when they offer Smarties with their café lattes. Of course, my kids love them; they are a memorable moment and add to their delight of going there. They increase the pester power. But it's diffi-cult to quantify this experience. It's an indirect effect, it colours the expe-rience and is part and parcel of my kids and hence our pleasure. Hence it is unquantifiably important.

But the point is the kid's reaction shows an exaptation. And we could use this! Perhaps we could focus a bit more on wowing the kids.

When we observe behaviour and listen to narrative sometimes we find things perhaps happen not so much 'on purpose' then as 'on repurpose' (the exaptation)! Rather like the way the feathers on a dinosaur used for warmth under pressure of evolution were repurposed to create flight. Or Dr Martens' boots were repurposed as a fashion item.

Hence, being aware of dispositional changes acts as an alert to how things are going and the need for a proactive response. And they can be used for segmentation purposes; for instance, in the flow of narratives, look to see where groups are coalescing.

Critically then, narrative research, unlike other forms, requires that we don't reject outliers and hunches because these show where things are or might be heading based on our own human understanding of the data, something a machine cannot do because it is incapable of embedding such human reasoning.

For instance, consider how Jeff Bezos of Amazon redesigned how books and CDs were packaged in order to make them easier to open. Although this idea was based on only one comment by a consumer, he immediately understood its importance and the opportunity was taken. No need for big data, just the need to understand the customer and be able to spot the great ideas that add value to the experience the customer has.

Which is why employees are so important to our understanding of the experience and how we can innovate.

In pitching narrative research on my website www.allaboutexperience.co.uk, the difference from evaluative techniques was noted. But Olaf Hermans also spotted the key issue that not all decisioning can be put into this frame.

Here is the Q&A between Olaf and Dave Snowden who originated this complexity thinking and its business application with his company, Cognitive Edge:

Olaf Hermans:

> This is very interesting stuff, Steven! The complexity view and the sense-making/interpretative approach based on narratives to predict (changes in) attitudes and behavior is a powerful alternative to the traditional evaluative methods. Still, it would be unwise to exclusively see man as an 'interpreting intelligence.' Affect, evaluation, as well as conditioned and planned behavior are other human 'modes' which may be salient under varying circumstances and which are relatively independent from the narrative mode. Decision making and behavior is more than trying to shape a meaningful and sensemaking next chapter to an evolving story of past events. Nevertheless, this is very very powerful as it can be analyzed with new technologies and as customers can be primed into a narrative mode (as they can be primed into e.g. an evaluative mode).

Dave Snowden:

I'm not sure that we see (wo)man as an 'interpreting intelligence', one of the key aspects of the approach is that we capture observations and micro-narratives (not really stories) and the high abstraction signification is descriptive not evaluative. Now interpretation sits somewhere between description and evaluation.

Decision making is formed by our individual and collective narratives of the past and present but yes of course there are aspects of autonomic and novelty receptive cognitive processing. But I would argue that the nature of those (and other factors) is revealed through the signification process (which adds to, rather than interpreting the original narrative material).

We also run mass sensing of networks to support decision making.

Customer Experience Research Method 2: Effective Journey Mapping

I often read articles on how to do journey mapping and however good they are at showing pretty pictures and process maps they don't cut the mustard with me.

Firstly, journey maps are a design framework. Their intent is to come up with the ideas that move the dial in terms of 'the experience the customer has.' So no budget to trial and test these ideas, no journey map.

But beyond that, a rush to process, a desire to draw boxes and get our internal stakeholders only into a room with Post-it notes feels not just wrong but extremely myopic and biased. Why? Because what I find is that firms frequently use maps as a schematic to think about physical processes and how they can industralise and scale them for firm efficiency. When in fact they should be used to consider the experience the customers has (think, feel, do) and how we can improve it! At least, that is, from a CX perspective.

Journey Maps also tend to operate in a linear way with firms, which is wrong. For while firms think linearly, customers do not, instead experiencing non-linear journeys where they do not think of every step and often relate to other so-called journeys in their recall (such as in the billing example). So ultimately, while I agree we do need to think about the linear physical layer in execution, we must start with understanding the customer and other stakeholders view. After all, if we have no view of the customer experience (outside–in) we risk making false assumptions and assume that say electricity and gas experience is just about a smartmeter or customers are really, really concerned with the same granular functional things as businesses are.

A physical-only approach will also miss out on the 50% of the experience that is about simple communication, things that can be easily resolved such as how easy it is to find that discount on the page.

The best way to do a journey map then is not to box and constrain it, but in my opinion simply use a blank sheet of paper and listen. In that way we leave ourselves open, unconstrained and with the capability not to impose an internal process logic.

After all, if experience is about 'seeing the world as the customer does' rather than an objective series of boxes, then point 1 must be to realise that customers are not calculating machines.

They do not 'sum all touchpoints.'

They do not have a linear journey through your experience.

They do not notice all the granular details that are important to you.

They do notice things that are personal and intangible and not part of a physical map.

And … most of the time, they just couldn't care less.

This is the challenge then. To listen to the customer's journey with you requires a system that is open and nonlinear, one that prevents us rushing to internal process first without considering customer benefit. And lets the customer journey emerge!

I also believe that a focus on 'analytical equity' – a process map, a statistical chart – uses a different part of the brain than that required for 'creative equity', which is unconstrained, inventive, nonlinear and requires a space to operate. And I would argue that it's creative equity that is important for good customer experience.

So with a blank sheet of paper and a budget how should we listen?

Here are a series of multimethod approaches I would apply to fill out the journey map with ideas based on open listening:

Journey Mapping Method 1: Don't Set the Purpose First

Purpose is essential, but I don't believe we should set the purpose of a journey map first for the simple reason that we immediately constrain things if we do. Of course, starting with a broad general fixed defining statement to help our experience design, such as 'trust our bank' or 'we care' is reasonable and I have no problem with having a trust or care measure on some dashboard (as long as we don't confuse that with the feelings related to trust or care); it's just that, what happens next in journey mapping? That's my concern.

Should we start measuring the drivers and destroyers of trust and care before we do anything else?

Should we only look at 'trust or care clues' and omit other states like being efficient or a feeling of pleasure that might help create a 'sense of trust and care'?

Do we really believe we are correct in the first place? Or are we drinking our own Kool-Aid?

Ultimately, we need purpose I agree but when we start to journey map, let's hold that back a little.

Indeed I have found that on reflection sometimes what we think we should do is in fact not what we should do! And prior to listening all we were doing was fit the data to our own prejudices.

So back to our Charles Clinkard example; they have a brilliant purpose focused on staff knowing how to fit kids' shoes. We go there because of it. But it is good enough already. Instead, what I observe 'in the behaviour' is how kids love to view the small in-store fish tank and the pain of the long wait time!

Now, this has nothing to do with staff excellence, but these things are important to whether we go there again and affects our feelings towards them. Kids will pester us after all, and you know I quite like seeing the fish as well, while I'm waiting for the long queue to clear.

The only thing is, they have failed to capture this source of value precisely because they have only focused on the quality of shoe fitters. They have rushed to map in the purpose before considering the broader picture

So Charles Clinkard could do a lot more! Perhaps they could make it three times bigger with fish-themed shoes and a fish app. And observing how people hate to wait they could find better ways to handle busy hours.

Likewise if Charles Clinkard had rushed to state some general emotion as their purpose, they might equally have missed these other emotional values – goal states you should be looking for – around fun and a sense of convenience!

In the same way I have seen stores such as the North Face sell adventure clothing equipment and fail to use the theme of adventure; and supermarkets sell sweets and fail to use the theme of the old-fashioned sweet shop in their one-size-fits-all experience, probably again because in terms of purpose they only see the functional. They only look from 'their' point of view.

In addition I might add that many companies highly underestimate the importance of the human side of the experience. How staff interaction is frequently the most memorable area, where we get most of our information from. Purpose clearly needs to encompass this as well.

So with our blank sheet of paper how should we populate it with ideas? Here I go through six methods to create some great ideas.

Ideation Method 1: Mine the Creative Equity

Don't just focus on 'as is' analytical equity: look to use social communities, research and engagement to collect ideas for continuous improvement from your suppliers, employees, customers, noncustomers (inside and outside your industry) and experts. You need to 'mine' your creative equity!

To do this build cross-silo innovation platforms, develop an innovation culture and critically a proper governance structure to turns ideas into action. One company that does this well is LVE, winners of the UK CEM Awards 2015. Here ideas generation is considered part of daily work and culture, taken seriously and invested in. Likewise, use social media for ideation; look at how Starbucks and Giff Gaff mine their creative equity.

Ideation Method 2: Gain Stakeholder Engagement and Ideas

Because journey mapping is a design framework we should first of all always work with a cross-silo stakeholder group involving senior decision makers.

Otherwise we risk coming up with some great ideas and no budget or traction. In addition, stakeholder involvement should involve other departments, suppliers, customers and where relevant other companies. After all, an alternative view is always ideal to define the future.

For instance, one critical stakeholder is the CIO. I mean they might know how we can use developing technologies such as AI to augment our journeys, or a single view of the customer.

Then once we have our stakeholder group, I have found it useful to immerse them in the data and get them to critically observe and engage the experience: get muddled in order to find a way to simplification and a new way that encompasses all significant angles and sources of value as follows: E×Q = f(PQ, SPQ, CJQ), based on drives as well as of course efficiency and excellence,

Ideation Method 3: Use Immersion

Immersion here means observation and walking the experience, both the one under investigation and stellar comparator experiences.

The benefit of this is twofold:

Firstly, it uncovers granular, in the moment but driving moments that customers might have forgotten; particularly important because customers cannot remember everything even if it is important to their bigger 'sense' of the experience.

One form of immersion I have seen for instance is to hot-house respondents, uncovering in the process the hidden experience. This is something researchers Mesh Experience are expert at: across multiple studies, Mesh has found that peer observation, seeing someone drinking a soft drink, eating chocolate or using their mobile phone, significantly affects brand consideration. These experiences largely go unmeasured by marketers yet they can have a big influence on people's perceptions of brands and, ultimately, sale.

Another form is how Innovation Bubble used behavioural psychology experts to walk the experience, using their psychology frame of reference to understand the deeper drives and having the eye to observe human behaviour. One interesting angle is how they use emotional landscaping to look for those conscious and nonconscious goal states that lie behind memory.

Secondly, immersion means creative thinking, seeing how you could engage new ways of doing things, creating new drives 'big' and 'small'.

The implication of immersion is that mapping does not require everything to be empirically grounded from a survey. The expert judgment of say an Alex Polizzi is just that, but it still has the power to absolutely change the experience.

Ideation Method 4: Engage Narrative Research

Apart from engaging expertise and using immersion, we also need to use data. So our mapping uses sources such as business intelligence, complaints, existing surveys, social media conversations and narrative research as already described. And remember that the journey map is a living, not a static document. In addition, 'do' analytics is critical. Understanding the physical actions customers perform with the why they perform it is a very powerful method that enables us to uncover CX opportunity: sometimes we don't need to survey everyone!

Ideation Method 5: Uncover Deeper and Hidden Drives

And now we go deeper still.

Another key method is to uncover deeper drives through depth interviewing. Here we use techniques such as rep grid which asks several 'And what does this mean for you' questions; the idea, to uncover the hidden drives in an experience that can set the design theme, such as we saw with Zappos (we care) and Starbucks (the Third Place).

This is for instance how Cranfield School of Management working with the London Symphony Orchestra uncovered new ways to differentiate the experience. Here customers tended to say on the surface, 'We want to see a stellar performance,' but in reality their additional deeper but hidden drives were 'to support their performers.' This sense of social identity led the orchestra to get the performers to socialise with the audience after the concert.

This is also how with a ferry line the in-port not just the ship experience was found to be important; here the ferry line could have created an 'emirates' of the ferry industry with an end-to-end experience that made money, for instance with great port facilities where people spent their money, if only they had thought beyond the immediate ROI.

Or how about Emirates themselves: they use small touches to change the look and feel of the experience. Here the food looks like food, the entertainment system is best practice and the employees are friendly. Human design, looking at what issues might drive a sense of overall care, at least compared to other airlines, is useful in such a commoditised industry as flying.

And once again in flying there is Delta Airlines. Hence we find an airline producing wonderful safety videos, normally a boring experience, and a focus on the fine details of turnaround time. To date they are also experimenting with preloading of hand luggage to save boarding time.

In all cases, the aim is to reach out to define differentiating experiences based on the customer's deeper drives beyond price and service. And one simple way to do this is to put in a separate drives layer on your journey map: one that identifies the compelling goal and subgoal states (conscious, nonconscious and affect-driven) that lead to and impact upon a behavioural response; whether that relates to a decision moment in the experience or outside the experience.

For myself, a key add-on approach is therefore to map a goal hierarchy, one that encompasses not just the main customer goals and subgoals but the sufficiency, necessary, facilitative and inhibitory links between goals. Surely an improvement on SERVQUAL which does not consider customer drives at all! (reference Clore, Ortony and Collins, The Cognitive Structure of Emotions).

Ideation Method 6: Uncover Nonconscious Response

To some this may sound a little strange, but it's not. In the decision moment, a lot of our decisions are made in this way. For example the classic moment when we find that we have arrived home after a car journey but our mind has been elsewhere: somehow all those difficult mechanical decisions have been made if not on autopilot then certainly without engaging our conscious self too much.

But nonconscious response is also apparent in things we think we can give a logical answer. How we can put lots of parameters on a purchasing decision so it appears logical, but in fact we have decided to go with one supplier already because they seem more trustworthy.

Or how what drives us emotionally is frequently derived from nonconscious influences that are not expressed or under-weighted in a survey. So I might tell you I am going to Caribou Coffee for a drink, but I really like the ambience and the fact that it's different from Starbucks!

Even in B2B we see this in the example of how CTOs buy on the basis of a nice-looking IT architecture diagram. Of course post hoc they will 100% pick up on the rational information that supports their case and ignore the rest.

This is why I like the approach of Emotix from Innovation Bubble. Using observation, immersion and data to uncover what customers are not so keen to express or cannot express. In this way we gain a deeper insight into the emotional nonconscious takeaway, and it allows us to re-emphasise the importance of some ideas over others, such as the impact of packaging and brand logos.

For instance, consider how we may say one thing but not only do another (observed behaviour) but have hidden unexpressed feelings. So 'Luton Airport' is convenient but it feels a 'grey experience', one I would not engage with. Or how employees feel customers are like this (elite) but customers themselves express themselves differently: an interpersonal gap in emotional understanding.

Filling Out the Framework

From these multiple listening methods we then derive a series of design ideas and strategies to enact. And because we have engaged stakeholders from the beginning we should be in a great position to understand what to do and to let our purpose and Journey Maps emerge.

It's here too that we start to get real, to use frameworks like Cynefin which help define whether an action is in the complicated, complex or other domain: all of which holds implications for investment and execution, and makes apparent the cultural and governance challenges to any design idea.

In this way we realise the benefits of going beyond fixing the basics, which is of great value because the thing which disappoints me most about journey mapping is the belief that 'fix' is the first thing we should do, followed by some linear process to reach the nirvana of delight. This is not wrong but risks efficiency myopia, ending up in an Opex reduction rather than a customer benefit result. Remember, there are plenty of ideas which you can come up with in parallel to fix that don't cost the earth and are essential to delivering the branded purpose.

Nonlinear Journey Mapping

Filling out the framework also means we view customer journeys/stories as being nonlinear. So, although a process map is linear, what we find by doing journey mapping in a nonlinear way is consider how experiences can:

- Bleed into each other – *billing affects network perception.*
- Be influenced by other aspects 'not in the experience' – *how Amazon influences perceptions of website performance.*
- Be affected by brand, social and industry prejudice before the zero moment of truth – *how the banking industry has a bad rap.*
- Become salient through peripheral clues, personal interaction and goal states – *which often drive emotional value and fleeting affects.*
- Be affected down the line by how things are socialised – *what is in fashion*, what we remember.
- Pick up on the importance of anticipation just before a 'touchpoint' – *the moment of anxiety before the plane lands.*
- Depend on goals, subgoals and nonconscious interaction – *what we need.*
- Are weighted by the decision moment not some abstract moment in the relationship – *it's all about what happens when we decide, not some post hoc rationalisation.*
- Show that transactional scores of a hygienic event are in fact unimportant – *web download speed is hygiene.*
- Show how unconsidered touchpoints or touchpoints that are fairly general and journey based are important – *such as a sense of confusion.*

In essence, nonlinear journey maps show us the creative potential of customer experience and challenge our notions of what is a touchpoint and what the important moments of truth are.

In addition, I have also found it useful to understand what corporate stakeholders themselves think their customers would say about them. Indeed, by comparing the results of our journey map and our stakeholder map we often find some key issues. For instance, with a large banking firm, I uncovered how the perceptions of their customers as elite were out of sync with their actual customer base.

Inside–Out Layer

Service Blueprint

Now armed with a clear view of brand identity, 'to be' experience and what we can realistically do, we can set our internal process map (called service blueprint) and look at the physical assets we need tailored to the experience we want to deliver.

So we can look for IT systems that match our ideation and vice versa; we may temper our 'to be' experience by what we know from a technology point of view; after all there is no reason why inside–out we can't add in experience innovations.

Critically, because we are now more grounded, we can adequately define the purpose and how to execute it. We can start to do 'experience design' building out, for instance, UX into CX!

Experience Design

And what does that mean? Well, we should be ready to execute fast prototyping and turn abstract ideas into practical reality by using service design houses or developing our own internal design function.

In my own work, I have even done this in a different and cheaper way by using the power of storytelling and video (remember the Pixar process) to help engage stakeholders and test out ideas. Then once we have developed the concept, set up test areas and pilots with an emphasis on engaging in design that goes beyond a single silo.

Hence, the emphasis in Experience design is to remain grounded on the theme and purpose, which means we end up with a shared endeavour with shared metrics of success. Designing, if you like for the journey not the silo and hence setting the stage for collaboration, a difficult thing when departments are usually only concerned with their own targets and mindset, as I saw, when I tried using social media to get a network operations centre to engage with customer perception data and collaborate with customer care!

Focus on Do

Finally, we give space to experience by enabling the trial and test to determine if we are setting our brand story and direction of travel correctly.

Here is an example:

A loyalty points company had a churn problem. They knew the 'drivers' to churn but still found that even with a churn analytics programme in place, their results were going down. What I did in this circumstance was to focus the business not just on this 'control negatives side of the house' but also on the larger aspects of the 'experience'.

So, instead of trying to reduce churn rates by 1% let's look at the bigger issue of how inert their customer base is (how demotivated they are!) and how we can motivate that group of consumers. After all a driver analysis of an inert base just comes up with no drivers to spend!

And this meant looking beyond the data and immediate root causes, pursuing instead a focus on what new things we could do by looking at not so much the drivers of churn but how customers were being motivated.

And what did we find when we engaged the experience and looked at customer drives? Well, there were numerous things such as how we could improve the website content, make it look less ugly and make the experience of buying a flight ticket less confusing.

And then there were the bigger themes, 'the sense of the experience.' Here customers spoke of it as being 'complicated'. That it was too youth orientated, an interesting theme because the company had thought of themselves as innovative, a policy that had alienated their VIP, more elderly customers.

And of course, a lot of these issues and many more related back to the identity of the company. What drives the company!

Here, they thought of themselves as a bank because customers redeem loyalty points. And much of the corporate stories reflected this, such as how they judged themselves against a supermarket coupons company rather than a company that sells 'travel'.

So my first Focus on Do was about inside–out mindset change: something missed if we had rushed to process efficiency! Here then I showed the firm that customers don't think like you, about the collection of points, they think of the benefit, the holiday.

You are a travel company, not a bank!

And thinking of your company like that immediately changes your identity, who you compare yourself to and what experiences you want to provide in the marketplace. So, the question then becomes, how could we make people feel like this company was an 'experience', one that reflects their interest in travel?

Which led to design ideas around gamification, the employee experience, a more inspirational look and feel to the website, a different comparison base – more Virgin Atlantic than NatWest Bank – alongside a careful look through the journey end to end and where drives could be applied; for instance, it was found that many of the experience issues and opportunities were down to simple communication! And changes here were not costly to make. We also found a stellar benchmark in how best to interact with customers in TUI.

Of course these ideas, generated from mapping and ideating 'driving experiences' would have been lost if we had obsessed about efficiency, ignored our internal capability to do CX and not let purpose emerge and just went with 'We're a bank, trust us.'

Finally, in this experience we engaged our cross-silo team, set up before we started the programme, to sign off the ideas to action. And this is important! For instance in another example, an airline set up a virtual cross-functional design team with IT. It was the job of this group to sign off ideas and obtain budget. In other words, they got their key internal stakeholders engaged in order to stop governance being a blockage to design-led action and pilot testing, which is the stage where we learn, adapt and determine ROI against metrics relevant to the action!

So in summary it seems to me that managing subjective response is more of a design and collaborative process where we need to engage expert hunches and creative ideas. Otherwise we risk closing our minds to opportunity and constraining ourselves to what is, which might not be where we could be.

A Final Note on Journey Mapping: *This New Way of Mapping Delivers a Better Understanding of Value*

Many companies believe that customers judge the value of their CX only through a process of summing up their touchpoint quality expectations. And that this process arises in a linear, granular, objective and efficiency focused way. But for me this is myopic because:

1. What is of value is frequently about the qualitative appeal. Something creative and new not just about prior expectations.
2. What is of value is salient and felt based on goals and sub-goals; it is not just about efficiency.
3. What is of value is often not just about touchpoints but more the overall *feel* (as in the statement, they are confusing to deal with).
4. What is of value is something not necessarily fixed but changes through time, is modulated and occurs across the customer story (a preferable word to journey)!
5. What is of value are peripheral and nonconscious clues.
6. What is of value are experiences that affect *how we feel as people*, that is, how a pleasant interaction lifts our mood: hence a cup of tea at the hairdressers.
7. What is of value is altered through the social environment.
8. And customers have a non-linear journey, where they do not sum up touchpoints except in a limited way.

Our value judgments are therefore moderated, mediated and directly affected by experience, and yes efficiency in context is important! It's just that, unfortunately, in our use of rational research the *tail wags the dog*.

Rather like in the early days of planet hunting, we only saw Jupiter-sized planets due to the limitations and bias of measurement, so in journey mapping, if we use a granular, efficiency approach we will see some big effects for sure, but miss out the bigger deal.

We Need a Better Way to Map, So We Can Come Up with Better Ways to Design CX

Beautiful Design of the Familiar

Finally I would like to raise some examples of what great CX design can do. And I don't mean major CAPEX investments; we are not talking building a Dubai shopping mall. I mean the many well-designed small experiences that hit on a customer drive, the beautiful design of the familiar, the kind of thing that makes us feel that something has been designed with us in mind, something that makes it an 'experience'.

One of the ways I like to find them is to walk the experience and actively look for and observe what drives the customer. Things customers do not receive at the moment, but could motivate, differentiate and make you money.

So on a plane flying from Dubai to London, I thought of how annoying the passenger lights are: these are the lights that you put on yourself to read a book. As they come from the top they affect other customers' space; not great when you want to go to sleep. Or how you want to hang your coat up rather than put it scrunched up in the overhead locker. It would not be too much effort to simply add a plastic hook in with the magazines or redesign the plastic food tray bolt.

And once the motivating moment is defined, we can then execute a better solution.

The following then are examples of drives: none of these would have been found through driver analytics, but are nonetheless influential and affect the customer's sense of the brand, at least when done well as an overall endeavour based on the theme/purpose rather than a piecemeal activity.

- Trunkis redesigned the standard suitcase by making them fun for kids.
- Gatwick Airport has a section of the airport which provides bird sounds to give a sense of a woodland walk.
- The Leela Hotel uses lobby space and classical Indian music to effect a relaxing mood.
- Qantas provides proper food in economy class. (This for me is memorable!)
- The Doubletree Hotel gives out hot baked cookies.
- Free pens are provided in Metro Bank and a nice blue colour merchandising and interior design to make us to feel warm towards the bank.
- A bank helps kids learn code, to make them feel they are more trustworthy as a bank through their engagement with the community.

Beautiful design of the familiar also makes 'experience the marketing'. Here is a great example.

Emirates offers the A380 experience in the Dubai Mall and London. This allows customers to have a go in a flight simulator, 'flying' Emirates on a route of their choice. What is the ROI of this? Difficult to say; no one would say, 'I will fly Emirates because of the simulator,' although you could see the effect over time after putting in place this 'experience'. But the key thing is, it gives a good feeling about Emirates to a broad audience. If we just focus on 'root cause, as is' effects, we will never go down this path.

Beautiful design is also how we build commitment through the operational relationship: in other words how the firm is seen as a person. For instance, if we just upsell and cross-sell then our sales figures in the short term may rise, but we are not seen to be in a good relationship with the customer. Likewise if the quality of our written communications is poor we will also be seen in this light.

However, if we show empathy and care in communications, this, I suggest, has much higher resonance in the mind of the consumer.

And, don't forget the employee experience! Apply the same CX techniques to employees. In fact one of the opportunities is how we could use employee experience as a differentiator.

Are you embedding the voice of the employee in your customer experience designs?

Management Implications

1. Researchers need to stop thinking like engineers and mathematicians; customers don't!
2. No customer walks out of a store saying that was a great 8.5 out of 10 experience; what they do is tell or recall a narrative. Narratives are more valid to how subjectivity works and inform us of how things are changing, picking up small clues and dispositions in the experience.
3. Where possible get customers themselves not a text algorithm to quantify their narrative.
4. Journey maps are a design framework. Their intent is to come up with the ideas that move the dial in terms of 'the experience the customer has.' So no budget to trial and test these ideas, no journey map.
5. The best way to do a journey map then is not to box and constrain it, but in my opinion simply use a blank sheet of paper and listen. Critically, let the map EMERGE!
6. I don't believe we should set the purpose of a journey map *first* for the simple reason that we immediately constrain things if we do.
7. Because journey mapping is a design framework we should always work with a cross-silo stakeholder group involving senior decision makers.
8. Immersion means observation and walking the experience, both the one under investigation and stellar comparator experiences.
9. Uncover deeper drives through depth interviewing.
10. Map a goal hierarchy – a vital addition to current journey maps.
11. Uncover nonconscious response.
12. Use behavioural psychology.
13. From multiple listening methods derive a series of design ideas and strategies to enact.
14. With a clear view of brand identity, 'to be experience' and what we can realistically do, we can set our internal process map (called service blueprint) and look at the physical assets we need tailored to the experience we want to deliver.
15. We give space to experience by enabling the trial and test to determine if we are setting our brand story and direction of travel correctly.

16. We define ROI through trial and test.
17. Beautiful design of the familiar does not mean large CAPEX investment!
18. Embed the customer experience techniques described in this book, because these are valid to the unit of measurement, the human mind.
19. Understand that CX enables options! 'What's the ROI of the font type of a website' is a crazy question but without it there is no website.
20. Don't get restricted and constrained through process efficiency alone; you need innovation space, to cocreate and embed knowledge in your offerings through a collaborative approach and utilisation of creative equity.
21. Drive through the theme/purpose of your business in the end-to-end journey: you need to be flexible but you must also be deliberate and consistent.

Hence, I challenge the process-orientated quantitative-only way mapping is done today and put forward the view that complexity and goal state theory have already devised frameworks we can use that embed the flowing not static 'experience the customer has,' not forgetting that this means developing our internal empathy to understand what it is to be a customer and to enable collaboration and an innovation platform to execute cross-silo change.

Part IV

Emotions and the Customer Experience

And now we get to emotion!

Companies commit a fatal flaw if they fail to account for emotion. After all, customer decision making cannot do without it.

Consider for instance the scientific experiments of leading neuroscientist Professor Antonio Damasio. Here patients who had their emotion centres damaged were found to be unable to make a decision. Because 'emotions put the body in the loop of reason' patients were unable to understand what was correct for them and unable to use the heuristics of emotion and perception to understand how to react in a complex situation.

Which is why I find the usual pushback – what's the ROI of emotional response – unacceptable, because the answer is in every decision you make! Emotion is not a condition underlying just NPS, CSAT, CES, in the moment text verbatim or that slippery concept loyalty. All decisions, even functional ones, such as buying a lightbulb, engage emotions!

Which means of course it is useless for us to say let's create this emotion because there are 101 things that a feeling could relate to.

So faced with this what should we do. Well I say focus not so much on emotion as on understanding emotional *value*. Defining where we can or could differentiate *the experience the customer has* based on these frequently hidden or underused emotional drives to use and buy.

So a brand such as Amex would not say 'Let's create a feeling of excitement'; they would say, 'How can we deliver more emotional *value* to our VIP customers'; leading ultimately to the titanium card and rumours of a card without limit!

However, because extracting emotional value relates to deeper drives, companies need to go beyond the use of traditional analytics and management approaches using 'agile', narrative and implicit techniques (e.g., A/B testing the emotion dimension) as well as engaging in experience design trial and test as we have seen.

Only then will we be able to create a compelling need, from which emotions emerge.

But of course, don't let me stop you being an artist. Saying we want to create emotion X from this experience is a great way to think and design. Just ensure you link it back to the concrete and understand that emotional response is multiplicit and frequently fleeting. If we want customers to say, 'If I go there, then I will feel I can trust that bank more,' we should have an image of what that sense of trust relates to.

It's not the emotions that matter; it's what they mean.

10

The Value of Emotions

Introduction

Emotions, that's fluffy data isn't it?

Well, I wouldn't say so much fluffy as different! After all, their importance can be seen in how you answer the following two questions:

Do you believe customers have emotions?

Do you believe customer's emotions affect how they make decisions?

If you answered yes to both questions then I am guessing you should be interested in how they work.

This means that if we are serious about emotions, we should accept that as subjective data they work differently from objective. Hence, we should not force emotion to behave like objective data.

Nor exclude emotion and miss out the importance of 'emotionally' important events, just because they are difficult to define! Otherwise we risk excluding the customer and focusing only on functional hygienic things under the mistaken belief that objective efficiency is the only thing that affects purchase.

Or worse, assume the functional is all that equates to the emotional!

Emotions Matter!

So let's be clear. Emotions matter! Without them customers wouldn't know how they felt or even better how they are 'about to feel' about an experience; and if they didn't know that they wouldn't know what is rational to them and how to make a decision!

To quote neuroscientist Antonio Damasio, emotions it seems, 'are not a luxury, they are essential to rational thinking and to normal social behaviour.' They 'put the body in the loop of reason.' They are affected by who we are: our personality, our drives, our mood that day and affect who we become and how we behave.

So, every time we are about to make a decision, we make it 'with emotion' whether we are conscious of it or not. For instance, when we are going to buy a shirt, check in at an airport, make a business transaction, visit a website, consider the trustworthiness of a call centre rep or the level of customer care in a hotel, emotions affect our assumed rationality, just as rationality affects our assumed emotionality.

Emotions are therefore a currency. They act as an informational feedback loop, mostly indirect, that tells us what kinds of options we have and what to do. And this is particularly important when we think of the types of decisions we have to make.

For instance, consider the following scenario adapted from the emotion literature.

Which would you choose to do today: go for a walk, learn a new piano piece or read a new book? Only you can make that decision! It would be impossible for a computer! Hence, emotions are required.

And here are some more examples where emotions help us:

In response to a decision-making moment where we have insufficient time to think about the risk: Emotions heighten our state of alertness moving the body to act. So, if a tiger walked in the room, you wouldn't make a cost-benefit calculation, you'd feel the 'fear and run.'

In response to conditions of uncertainty: So in a political debate, we may not know what our politicians are talking about but the way they look, talk, hold themselves, how the press has reported their opinion, what feels socially right all has a bearing on who comes out best.

In response to conditions of 'assumed' certainty: Emotions highlight which way we should sway in a purchasing contract; noticing potential small clues that might present a future threat.

What we remember: Emotions help us to learn; they inform our emotional expectations. So, when I decide to go shopping in that store again I pull out of memory my emotional fears, such as will I have to queue a long time or will it be raining, as well as my emotional desires: ' If I go there, then I will feel…'

Hence, through semantic content (the phenomenology of specific emotion words), valence (positive and negative) and arousal (high and low stress) emotions work hand in glove 'with cognition' (both conscious and nonconscious), our personality, mood and social environment to help us know what is important. And behind it all is how we are primed: what *drives* the customer in different modes of decision making whether this is immediate or where we anticipate emotional reward (the former often leading to poorer decision making than the latter).

Hence we are not, as some companies think, rational cost-benefit calculators!

And emotions do affect behaviour: which should answer the ROI question!

And emotions' effect can be considerable. As long as it is considered in context.

For instance, as we have seen before a firm might believe customers always make decisions in a cost-benefit way. But in reality although customers might initially be interested in the price and functionality, emotionally when they go into a store they feel that this is not a nice environment and go and buy from a more expensive store.

And if emotions can undermine the firm's assumptions of how customers behave, so they can be used by more emotionally aware firms to help them make money.

Hence, at the design level restaurants apply mood lighting, sporting events build a sense of anticipation, marketing creates a social buzz and at least some of the better insurance companies communicate complex policies with a pleasing clarity.

So, if we fail to consider emotion in our CX strategy, we discount the human parameter and make suboptimal choices.

We need therefore, a framework.

The problem is traditional frameworks and survey approaches assume emotion is something objective not subjective! They treat emotion as they would a fixed, rational, root-cause-based engineering project: where you get a number, get an algorithm and everything works consistently.

But in this way, instead of measuring the personal we reduce it to some objective position outside ourselves. We assume our brains work 'like a calculative regression model' and in this way discount the very thing we are interested in measuring: what it means to *feel*. In essence, a rational survey also does NOT effectively measure an emotional response; even if you use emotional words. And anyway, Norbert teaches us that surveys out of context are poor representations of what goes on in the 'decision moment'.

So what to do?

Well I believe to extract emotional value we need to look at the characteristics of an emotional response: where it arises, how it arises (what deeper drives are leading to this type of response), and the management implications.

Emotional Characteristics 1: They Are *Fleeting*

Emotions are a cognitive agitation of the nervous system which means they are *fleeting* by nature. For instance, if something bad happens such as 'I can't find the phone number on the website' or 'That customer service guy at the airport was arrogant and rude,' a flash of emotion will come over me. This will guide and direct *with cognition* how I behave in that decision moment.

It is this fleetingness then that holds implications for how we extract emotional value, because now we have to be aware of when it happens and why, in other words, the relationship to my goals.

The problem is we cannot depend on a survey for this for the following reasons.

Fleetingness Is Missed

Fleetingness means that in a negative moment while my recommendation score may go down to 3 out of 10 within the hour once the diffuse chemicals of emotion subside and the issue is resolved the way I feel returns back to a more normal 8 out of 10.

In this way the traditional survey approach is none the wiser.

And what applies to negative emotion also applies to positive. So if I experience something new and exciting, then I can expect to feel an emotion. But once I have learnt what it's all about, don't be surprised if I feel things a little less the next time around, something called homeostatic regulation.[22]

Hence, if I depend on attitudinal survey response I might find that the first time I go to LUSH I score it 9 out of 10 and it emotionally means something: I am thrilled; the second time I again score it 9 out of 10 but this time it's less of an emotional response, more attitudinal. And the tenth time I go there, I still like it but I start to notice negative things! And score it 8 out of 10, because I care: in this way sometimes negative emotions correlate to a good attitude and positive spend!

Fleetingness Is Not the Whole Story

Fleeting emotional statements are therefore important, but in seeking to measure them we also find a few problems. Why? Because the fleeting statement needs unpicking – it is not the whole emotional story!

For instance, if we captured flashes of emotion through verbatim or social media commentary and applied a sentiment text algorithm, we might over-interpret our findings. So in reality customers say things fleetingly that sound bad but actually they feel good. Hence, standing in a queue for a new iPhone I might complain bitterly about the wait time and is it really going to be worth it. But I really love Apple; that's why I am prepared to wait. Or I might take 100 pieces of text all with the word bad in them, but none of them will tell me 'how' bad.

With emotion, text is only the surface layer, which is why customers need to title and self-score their own text in terms of emotional importance (Source: Cognitive Edge). If they can!

And of course this represents a challenge to enterprise feedback management and other traditional approaches that assume there must always be a root cause! But many times, as we have seen, not only do we find a disconnect between a statement and how customers feel, but also customers have a tendency to talk about the sense of things. Too often we throw this away as fluffy information when in fact it is fundamental.

Fleetingness Is Nonconscious

Fleetingness can also mean emotional effects are nonconscious; which means traditional research simply fails to report them.

Hence, as I work through a website, I feel unimpressed or distinctly bland about its look and feel. Its black on white scroll bar, amateur-looking text and poor pictures that do not inspire. But all this is not something I register consciously; it is an effect that influences perception but in an understated way.

How to Research Fleetingness

Fleetingness causes us to under-report, over-interpret and nonreport emotion! It also means we cannot reconstruct emotional feelings by getting customers to respond to some survey served up three weeks after the decision event.

Which make me wonder about the kind of surveys that do supposedly correlate an emotion word with behaviour; perhaps they are just conflating emotion with attitude! Sure customers may tick the box, 'I am very happy,' but that is just an alternative word for satisfied or recommend! And anyway, you are asking me outside an event to tell you how I felt in the event!

Hence, just putting emotion words on a scale and asking consumers to rate them is not the same as meaning they feel them!

For instance:

> Going back to that example of a high promoter market where I found that emotional intensity was profoundly positive: I also noted that one-third of promoters in part or in full stated they would never, ever recommend the company. Likewise on the detractor side, 65% did NOT state they would never, ever recommend the company and the emotional intensity was at a passive level.
> It seems that customers were responding based on their behaviour: 'I use this a lot; it must be good,' and because I say it's good I'll also tick the high emotional intensity score as well. Or I don't really think about this brand, but I don't know about alternatives, so it feels pretty bland, but because I use it a lot I'll tick that I am happy to recommend.
> And anyway, a 0–10 scale tells customers what we are after.
> I would also bet that you could exchange any set of emotion words with another set and still get a valid and reliable result.

I mean, surely if we were measuring emotion, we might see:

Less emotional intensity: I like this place; I know that; I don't have to feel it so intensely.

An emotion norm to the product or service in question: So Fabergé eggs score high in valence and intensity, electricity and gas less so 'but it doesn't mean anything bad.'

This is why in research I use alternative approaches that are designed for fleetingness: for instance, immersion, taking narratives from multiple perspectives – consumer, employee and experts – as they walk through the experience, research that focuses more on gut reaction than considered response, and more realtime measures that focus on getting responses 'in the decision moment.' Not forgetting of course that a 'decision moment' can be when we decide to go to a store not when we are in the store!

In addition, because narrative is so important for picking up emotions' fleeting semantic content, how it feels personally, I would embed this approach more than a scaled one while embedding emotional learning in our emotional journey maps (see Fig 10.1 for an example of approach).

Health case study 2: Smoking cessation

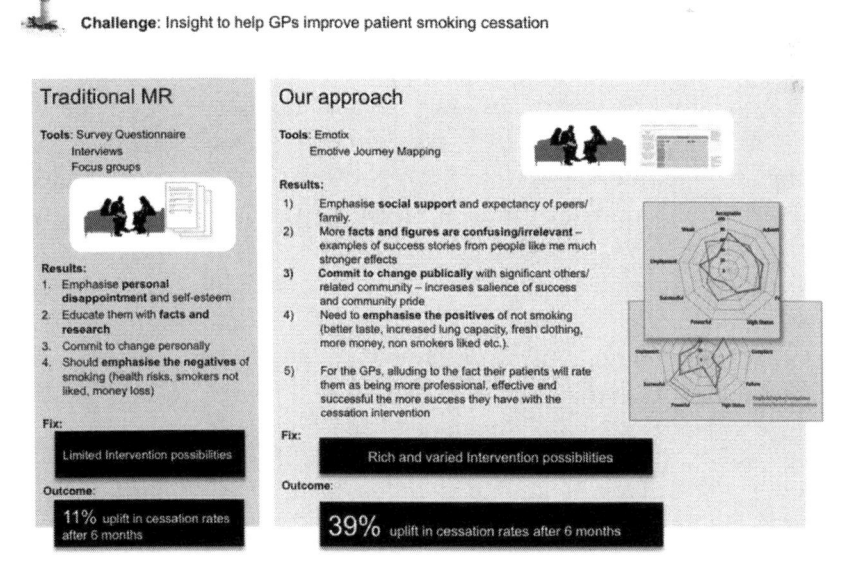

Challenge: Insight to help GPs improve patient smoking cessation

Traditional MR	Our approach
Tools: Survey Questionnaire, Interviews, Focus groups	**Tools:** Emotix, Emotive Journey Mapping
Results: 1. Emphasise **personal disappointment** and self-esteem 2. Educate them with **facts and research** 3. Commit to change personally 4. Should **emphasise the negatives** of smoking (health risks, smokers not liked, money loss)	**Results:** 1) Emphasise **social support** and expectancy of peers/family. 2) More **facts and figures are confusing/irrelevant** – examples of success stories from people like me much stronger effects 3) **Commit to change publically** with significant others/related community – increases salience of success and community pride 4) Need to **emphasise the positives** of not smoking (better taste, increased lung capacity, fresh clothing, more money, non smokers liked etc.). 5) For the GPs, alluding to the fact their patients will rate them as being more professional, effective and successful the more success they have with the cessation intervention
Fix: Limited Intervention possibilities	**Fix:** Rich and varied Intervention possibilities
Outcome: 11% uplift in cessation rates after 6 months	**Outcome:** 39% uplift in cessation rates after 6 months

Fig. 10.1 Emotix study

> Emotix © is our unique psychological engine - developed from the work of Greenwald (Implicit Attitude Theory, 1990's) and Kahneman (System 1 & System 2, decision making). The engine has been jointly developed by our neuropsychologists and psychometricians to reveal people's subconscious influences and associations that impact on their decisions and behaviour. As psychologists we know that non conscious emotions drive 80% of consumer decisions and behaviours.
>
> Unlike surveys, questionnaires and standard interviews it removes interviewee social bias and conformity and unlocks the 'true' influences and real needs that drive action and inaction. It measures both prevalence and strength of influence. It is an online test that produces response rates above 85% and that can be deployed globally with thousands of respondents to provide deep psychological market analysis.

Emotional Characteristic 2: They Inform and Guide Our *Dispositions*

Emotions act as 'weak signals'; they let us know what the 'experience' means and where our plans are heading; in other words, its disposition. They arise in the interruption of goal and subgoal states (Oatley). And this as we have seen is a concept very different from traditional notions of how emotions drive and destroy value in a fixed way.

So, if I phone up a call centre and find that the rep's tone of voice is warm and empathetic I am likely to feel this is a company worth doing business with. I can trust them for the long term. I react to the disposition of the moment!

Likewise, emotions alert us to information that justifies an existing predisposition. Hence, if I love Apple even if their call centre staff gave me poor quality service, my disposition is to forgive them.

Emotional disposition is also critical in determining levels of anticipated emotional reward. A week ago I had a bad experience at NatWest bank, due to their sales-driven rep who didn't consider the fact that they had called me in to do a simple transaction. Now, I anticipate that I'd get a better emotional reward 'if' I went to Metro Bank; the only problem being of course, the ease of moving.

Inertia and the perceived stress and effort of changing accounts clearly hold another emotional connotation that affects behaviour. And note how the level of emotional upset – due to this waste of my time and inability to listen to my needs – is strong enough for me to want to move, not something I'd forget.

In all these cases, then, it's this dispositional aspect of emotion that highlights the things I notice, remember, and affects my direction of travel. A disposition heavily influenced by what Dr Simon Moore calls peripheral clues. Clues you have to watch for before they reach a tipping point of change and undermine your resilience to market movements.

An example of this is how at one hotel I was asked to give a €20 deposit for a travel adaptor. The next day, when I asked for this back, a new person on reception refused, in fact by his body language and tone of voice he seemed to blame me for the lack of evidence on their inventory. So although I eventually got my €20 the emotional feeling of being blamed altered my feelings towards the hotel in ways that emphasised the more negative aspects of the experience. Negative aspects that made me feel a year later, 'This wasn't such a great place, I might have to be careful here' even though I have forgotten the bad event! This is something firms fail to realise: negative events may be forgotten or forgiven, but what's important is not necessarily the specific event but how this affects my relationship with the brand overall. And what I intend to do there. Will I browse more often? Will I treat it as a transaction?

So next year when I am looking for a hotel in the same area, I might find it easier to pick out of memory my feelings of blandness or negativity towards the hotel (even if the actual feeling per se is muted), which leads me to become predisposed to NOT go there because it is a long way from the city centre (I am less resilient to going elsewhere). Contrast this with a more friendly response, which might have led me towards a more positive predisposition: 'Even though it's a long way out, it's a great place to stay' (the resonant cues encourage me to overhaul my initial inertia to making the effort).

Dispositions can also be affected by the mood of the environment, which firms do have some control over. So, the Leela Hotel has a relaxing lobby environment, a memorable 'peak' moment that affects my ongoing relationship.

Personally I feel this mood impact is underrated. When we make a decision it comes with memory, 'If I go there, then I will feel....' But memory is also tied to visualisation. So when I think of this hotel, I also remember not just 'with emotion' but what the emotion relates to, namely the open relaxing lobby, water features and mood music.

In a very practical but difficult to quantify way, the Leela Hotel lobby moderates the negative impact of how I had one difficult night's sleep because of the loud TV from the next-door room.

Firms, then, need to watch out for dispositions through small clues in the experience because these can have big emotional and remembered effects. But unfortunately once again research and management processes are not well geared for this.

Especially where these are hidden, such as how a customer might say, 'I' go to Starbucks for a coffee; but her real drive and disposition is to sit down and talk to friends. Or they are modulators.

It is also worth remembering once again, that dispositions only give us options. My cognition may come into play and I override this; so even though I was not treated well last time, I'll go back to that hotel because it's 'convenient,' 'I know it,' and so forth. And of course after I say that, I will then feel that the bad event was just one of those things!

How to Research Disposition

Hence, if we are to engage in emotional design, we need to identify those dispositional trends both stated and hidden that are being guided or could be guided by our emotional reactions. That means we must go beyond traditional research approaches which only use complaints data, social media and surveys to measure emotion while missing its *fleeting*, hidden and dispositional nature.

For instance, traditional surveys:

- Focus only on root-cause effects when emotions are typically indirectly effectual and often nonconsciously felt
- Fail to pick up on the broader customer narrative (the modulators) or fleeting 'effects' as they happen
- Fail to account for outliers as weak signals
- Fail to understand if a score is resilient and meaningful or not
- Fail to research for further 'anticipated emotional reward' once I have had a chance to reflect cognitively on my emotion state towards say a bank in comparison to its competitors

However, as we have seen, there are better ways to measure emotional disposition using story metrics, the type of approach that collates self-signified narratives (sensemaker) and identifies how things are moving, their disposition, their flow. Another advantage of this, worth mentioning, is that to change the measure you would have to literally change the story, making it far more difficult to game.

And I also note how we can see the resilience of a score through this method: whether my giving you 9 out of 10 is something I mean or something just gifted; whether my 'within score' disposition is weak or strong, durable or not. Something immensely important and missing in datasets today! I mean, I give you a 9 but it may be a weakly felt 9, less committed, more prone to change – I churn and inertia – I am not engaged.

Consider this, if we could measure this 'resilience' we might find that what seems like a strong so-called promoter market is actually not so emotionally engaged. Or we might find that voters say they will support one party but are really not so sure.

So use the KPI and methodology behind 'more stories like this, fewer stories like that'.

But story metrics are not the only thing.

We can also consider using value-in-use frameworks to unpack hidden dispositions around anticipated emotional reward. For instance, the work Cranfield School of Management has done with rep grid; or how we might benchmark ourselves against experiences outside our industry, helping us define 'potential' emotional needs that might guide our disposition or use the trained eye of industry experts; or say behavioural psychology to determine the critical nonconscious emotional dispositions: either through immersion or quantitatively.

Establishing the Value of Emotion

So far then, we have seen how the fleeting, nonconscious, dispositional and anticipatory nature of emotion (how we reflect on emotion) requires different techniques of research and management. But even if we do pick up on emotion, how does it affect ROI?

Well, in the immediacy of a situation, emotions certainly have direct root-cause effects. We can say, 'We feel an emotion because of X,' as long as we get close enough to the decision-making moment to ask! Hence, if 'I' felt aggrieved by the jewellry shop that broke my watch in the repair shop then blamed me, 'I feel it,' 'I can say why,' I never go back there again.

But this root-cause effect of emotion on ROI is not necessarily the case every time! Many times a flash of emotion does not directly relate to our spending less money or using you less often, as we saw in the travel adapter and hotel lobby examples.

So, the fact that 'I' felt ripped off by having to pay £3 for a printing company to put my memory stick into their machine is not going to directly lead me to leave or use them less often.

In this way emotions also indirectly affect value.

Which leads us back to the concept of how an emergent sense of an experience informs our anticipation of an emotional reward: such as in the statement 'I trust that brand.' And this sense as we now know is built up from many small fleeting and dispositional emotions or affects we experience (these are the modulators). All things that are not so easy for traditional research to define because it assumes emotions act objectively and in a root-cause manner against a set scale.

Hence, we have to design in agile techniques of measurement and management. Otherwise, we will simply not be able to focus on emotion's effects and be unaware of how customers use emotions to make decisions through understanding what an experience means.

But ultimately the real value of emotions comes from how we can use them to identify a new emotional value (a *goal state*), which means a capability to take the risk. After all, if you add something new, you won't know if it works unless you try it!

Or if change management is difficult, at least augment existing experiences with emotional value looking for high impact experiences.

And the best 'Experience' brands know this. They understand that it's experiences that create emotions. It's just that defining the right experiences to amplify depends on identification from 'use' of those experiences that are currently emotionally important but ill-used or experiences that could provide a deeper emotional need repurposed from other experiences. Such as how Overbury took the concept of a concierge service into a construction environment.

Hence, they develop a deep intuitive connection to the customer. They develop their creative equity and empathy; they know what it is to 'live' the experience.

But for me I find even the word emotion problematic. We are not looking for an emotion; we are looking for more meaningful experiences that relate to what drives us as consumers, from which an emotion, as with an attitude, is derived.

Hence, I would argue we should use the term *drives* not emotion, because this encapsulates all aspects of customer behaviour, emotional, rational, nonconscious, conditioned, behavioural and it sets the frame for us to define what deeper drives we can uncover, how we can create an impulse to act, an effect, not just be efficient.

Not so much emotion as emotional thinking.

An Airport Story

To summarise, then, we can see emotional effects in the following story. Here, emotion is an outcome of a number of fleeting, root-cause, anticipatory and emergent events, many of which although not individually influential on ROI, taken together do build a sense of the experience that does affect how I remember and my behaviour next time around.

And managing the sense of experience through understanding drives of course brings us back to the importance of the theme and the purpose.

> Luton Airport has a bad reputation. Voted one of the worst airports in the United Kingdom, any customer would be predisposed to think ill of the experience. And predispositions tend to drive what we pay attention to! Hence, on 3 November 2015, my emotional expectations were not good. If I go there then I expect there will be long delays on the motorway, the airport to feel bland and unwelcoming, the planes to be delayed by fog and so forth.
>
> Initially this disposition was countered; the radio broadcast told me that the planes were leaving on time, and the journey on the motorway only had slight delays!
>
> But once in the airport things were rather different. Aesthetically the building and layout felt quite functional and although I was pleased with the lack of queues, and my early arrival, there were certainly some emotional flashpoints.
>
> First up was the inevitable need to buy the plastic bags. Why is this? They're free at Heathrow; this is just money-grubbing behaviour. Then the fact I couldn't find a £1 coin in my bag having given the last ones to the taxi driver. What next! A long queue at the ATM because two out of three of the machines were broken. OK there's no queue at the currency exchange; let's ask if I can exchange my coins for a one pound one whilst doing a currency transaction for euros.
>
> All well and good until a minor issue starts when after agreeing to this, she promptly forgets, hands the euro currency over and closes the till.
>
> Eventually, I get the coin, get the bag, realise I have brought over 100 ml of shaving foam and have to throw that away (memorable!) but like the fact I can easily get through security to departures.
>
> I soon forget the slight upset of the coin incident and the plastic bag; even though at the time my irritation was high.
>
> So, if you asked me 'What I remember,' here are the things I would say:
>
> The fact I had to throw away a personal item (the over 100-ml can of shaving foam) is one. Then there is the near argument with the currency rep. I think it would also be the long queues at the ATM because of the breakdown and the way I got around this. Then there are the £1 plastic bag

containers and the social sense from others that Luton is not a great airport. However, I have to say there are a number of other positive emotional needs that were satisfied. So, the convenience of getting to the airport, the fact there were few queues is critical. The fact that I avoid having to go to Heathrow! (Emotions are set in context.)

So personal loss would be the first thing, followed by emotionally difficult situations, especially in face-to-face situations; then some positive if muted, in this case, moments of ease.

But there is also another aspect to this. Visual memory is important; emotional memories come with a sense of place. So the bland aesthetic and functional atmosphere does nothing to amplify the good memories but certainly justifies the bad or bland! It seems even if we design it or not, there is always an emotional experience.

In all this I build a sense of whether this is a good place to do business in or not; I build a disposition: 'If I go there then I will feel' a certain sense. And because emotions are tied in to what we remember this is important for decision making.

But in addition, there are a lot of nonconscious emotional effects we have to consider, things that weight the experience.

And this is also important, I mean perhaps next time the emotional weighting remains with the convenient location and I don't care about the rest!

Hence, what I say on 19 December 2015 when I am thinking of going from Luton Airport again is: well, the emotional drive remains; it's a great location for me, I can avoid Heathrow.

So I would still use it.

But I am predisposed to use it functionally. I don't see any need to go early to engage with the brand, to shop or buy some food there. I also feel on my guard, the plastic bag purchase and the easyJet staff experience drill home to me that this is a functional place.

Hence, for Luton Airport the emotional takeaway although strongly correlated to value in terms of the functional (which is emotional! I do like convenience) exhibits no strong pull of emotional value that goes beyond that and towards engaging with the airport experience. The ExQ in the equation is clearly functional.

Now today perhaps things are sufficient for Luton Airport to just depend on functional needs. (I feel pleased because it's conveniently located, I don't have to go down to Heathrow and it has few queues at security – at least the last time I went), but resting on your laurels may leave you open to competitive threat and through failing to ask the emotional needs question, misses the potential of your experience to make higher margins.

And this is important.

For instance with one ferry line under pressure from a cheaper alternative, we found that there were plenty of emotional need factors they could compete on. They could for instance, make the onboarding experience, a common emotional flashpoint better, or they could innovate the port experience, making it more of a destination and design in signature ferry experiences for the family: don't just buy the cheapest; come and see our aquarium.

Likewise, we should not forget that 'the way we feel in the moment' before will also make us feel that way when we go back again. So, as a takeaway moment the plastic bags may not feel particularly important when I reflect back. But when I go back, it is a resonant moment I am attuned to notice and it becomes again impactful.

Designing for Emotions

No company has a probe into the consumer's head allowing them to manage emotions for profit!

Emotions, to state the obvious, are personal not controllable.

But this isn't a council of despair. Some of the most successful companies have successfully competed in this space. For instance, think how Apple, Zappos, Amazon and Overbury designed in 'emotional value' based on use (value-in-use), whether this is respectively a focus on inspiration, care, personalisation or a nonconfrontational environment.

Or even better, those companies that are simply 'more emotionally aware,' thinking through the implications of any interaction from the point of view of the customer, not just their sales target.

So how can we design for emotion? Well, here are a few strategies:

Identify Emotional Drives

One of the problems of traditional surveys is that they only scratch the surface of customer response; getting to know how people feel, their emotional drives, requires a deeper understanding.

For instance, how can we identify those micro-experiences that colour our experience but are frequently forgotten? Such as the way my mobile phone confuses me with the number of apps it holds, goes dark too quickly or fails to let me hear what is being said on the long mute message. Likewise, how one financial services company speaks to its clients too colloquially or fails to appreciate that I don't want the getting a mortgage experience to be effortless due to its effect on trust.

All these fleeting things act as points of friction and inertia that mask the positive benefits of the experience, things that can only be identified through observation and the type of customer experience research techniques we have identified earlier. If I could wrap this into one statement I would say: use consumer psychology.

Avoid Loss Aversion

The work of Kahneman on prospect theory is the most obvious starting point.

Here we see how customers are more sensitive to losses than gains hence defining psychological losses and how we can prevent these is important.

This is of course akin to fixing breakages in a physical environment, except that here we are looking for pain points in the psychological engagement a customer has with a brand. So, if I go to a website, I may say the look and feel is ugly, how the content is uninformative and complex in addition to the more physical aspects of slowness of speed.

Negative emotions thus leave 'emotional tags' (Vohs, Baumeister) or 'somatic markers' (Damasio) that as an organisation we need to be aware of. And these can come from anywhere in the customer journey. So 'banks are bad' is a 'tag' derived from the credit crunch.

In design therefore we have to be conscious of these and moderate them as well as be aware of the types of loss aversion. So, we need to be aware of the emotional journey, where sadness and disgust arise; points of stress, the fact that personal distress can create foolish risk and flashpoints between say a call centre and a patient phoning in with a suspected condition. Likewise, watch for how customers and employees use anticipated emotional outcomes and what this might mean in design.

So, with one ferry line, I used a heart monitor and self-report to identify how anticipated emotion was most important prior to arriving in port. It's at that moment that customers (and employees) are most aware of the stress of arrival. What are the instructions to get off the ferry? Will there be a mad rush? This is one of the reasons I suspect that easyJet no longer allows customers to sit anywhere, hence avoiding the distress of getting on the plane first. Clearly, managing stress is a key principle.

I would also be aware of counterfactual thinking. A bad negative moment will lead customers to ruminate on the consequences; learning 'not' to use you next time. Hence, service recovery and troubleshooting when things go wrong are critical elements, as is, once again, your interaction model and the debrief.

And loss aversion is also about *'proactive'* loss aversion.

So, being able to pick up a degraded cell on a network, and preventing it spreading or identifying 'ice ahead' on a road are excellent examples where the 'machine' can assist the emotional! And assist or augmentation is the optimum word.

In a similar way we may use social media or story metrics as a datafeed to proactively identify points of pain (and amplify points of pleasure) that others may well be experiencing, but just have not made the effort to report.

Be Personal

Emotions are personal and because of this, they open up the possibility of micro-segmenting, identifying slight differences between individuals that enable personalisation. And the big advantage of the digital economy is that the price of this personalisation is sufficiently low and scalable that we can start to see how we can economically justify such an approach. Mass customisation not mass production is now enabled! Are you ready for it?

Be Social

Emotions are affected by the social environment. How others see us, how we feel we should respond. So designing for social effect is important.

So the reason restaurants position customers close to the window is to achieve this effect. We can't see the empty restaurant behind but it feels a social place, somewhere people like to go which means it must be good.

Enhance the Mood

Our current emotional state affects our decisioning. The problem is there are very few levers by which a firm can influence the personal. However, there is one area that we do control, that is the physical or digital environment through which a consumer/employee engages.

Getting this right can therefore influence the right memories of an experience or moderate any negative ones.

Likewise, of course, we can get this badly wrong. So we may think that forcing customers to spend money by giving them a time limit on a digital experience is a good idea, but even though we see a rise in spend we are oblivious to the negative losses through the bad stressful feeling created.

Or how about the way we see sales rise through pushing out an endless stream of upsell emails, but again create a bad feeling about this method of engagement.

Mood effects also are important to consider because customers and employees can be put in the right frame of mind. And good mood enhancement means your stakeholders will seek information that is congruent to their mood!

Respond to Dispositional Characteristics

This is being aware and responding to how things change in the emotional environment, understanding what to amplify and what to remove, not waiting for something to happen: identify it early.

Hence, in one piece of activity, through story metrics and emotional intensity measures I identified that website perceptions were being impacted upon by a negative disposition towards rude customer service. Only a few people were affected, but the evidence demonstrated that it was affecting value and was a growing disposition.

Likewise, use of social media is good for picking up on emotional dispositions (often of the negative kind).

Account for the Nonconscious

Because so many of the emotions we experience are received in the moment and pass us by, a key design principle is to affect us emotionally at a nonconscious level, for instance, the look of an Apple store that spells innovation and technological capability. Or the fun way Virgin Atlantic would deliver service. We also see this in the principle of Nudge.

Critically, to understand nonconscious effects it is useful to immerse ourselves in 'what happens' and observe customer behaviour in the experience. This enables us to see how they use products and services and to define new sources of value, such as in the Charles Clinkard experience where I noticed kids' reactions to the aquarium.

Observation is of key importance to picking up nonconscious emotion and value: observations best done from multiple qualitative perspectives to reach a consensus based on expert judgement.

Create Memorable Moments

We should also create memorable effects; things that 'stick in the memory' and reinforce a good impression of the brand. For instance, the Harrods' shop window display at Christmas or the reasons why companies advertise on Formula 1 race cars. But memorable moments don't have to be capital intensive. For instance, the Smarties given out at a meal in a restaurant are a significant memorable moment for my kids which influences whether we go back there again.

Meatballs in Ikea! is another example.

But I would say what really counts in emotion is the use of the humanic – the person-to-person interaction. Hence, technology needs to augment and empower such interactions or if we are left with just a technology one, do our best to make it feel personalised.

Manage Stress

Whatever customers say on a survey before they go into an experience, in the moment, they can make different and potentially damaging decisions under stress. For instance, generating complexity in an environment where emotional expectations of ease are desired is liable to cause negative stress. Such as in complex insurance form design or when technology fails to work easily or when a single portal is provided for multiple applications leading to a refusal of the customer to engage.

For me at least as a nondigital native there is nothing more annoying than technology making my life more complex.

Alert and Understand Specific Emotions

I never understood why journey maps, sentiment analysis and social media report on emotion (assuming they measure an emotion that is!) in a positive–negative way. It's not such generalities that count but the specific meaning of an emotion. For instance, as humans we try to avoid feelings such as regret, or we look for reassurance.

Understanding the implications of an emotional feeling means we are in a better position to design for it; which is important when you consider how all decisions involve predictions of future feelings.

So why do we still report general feelings? For instance, it's far better for sentiment to report more specific words and alert us to strong emotions such as disgust even if it is only a few people reporting it; after all, if one person feels it there is an increased likelihood that others will but not report it, such is the way of managing dispositions. Look to alert and effect change dynamically.

Here is Dr Simon Moore's view on some of the usual ways we look at emotion:

> Start with what customers want to feel or think they are going to end up feeling – and don't treat emotions just as a 'positive' vs 'negative' that is archaic – if your angry or frustrated you are still connected – you still want to tell that brand – if you're bored or indifferent on the other hand then the brand has just lost you….

Measure Emotions Effectively

Seems obvious but if we are going to understand emotions in design it might be an idea to measure them! This need not be onerous but be aware that emotion items on a scale do not necessarily reflect an emotion felt, as we have seen with response bias and rationalised response. Likewise a sentiment or verbatim even if it expresses emotion words are also not necessarily representative of how people feel: I hate this queue outside Apple, but I love the brand. Attribution is important.

This is why takeaway emotions are best measured with techniques that include uncovering its nonconscious nature. If we avoid uncovering nonconscious (and conscious) emotion states then we will not understand what it is to feel.

Account for Heuristics and Bias

No brief would be complete without some accounting for the heuristics and biases of emotion, such as the Peak-end rule (although I note that Kahneman himself specifically avoids drawing CX conclusions from this; see Note 10 in his book *Well-being* where he references the criticality of goal states!) affect heuristic and numerous other responses which should be considered in an experience design. This can be a complex area hence I engage behavioural psychologists in looking at these effects.

Uncover What Drives the Customer

A lot has been said about this already, so it seems appropriate to end with the core principle. Emotions are based on how we are primed. Understand these and we will hold a better understanding of how to design. This is why some design houses are looking to personality-based segmentation schemas, and others towards customer-designed and personalised journeys. This is why I look to goal-state mapping.

Use Goal State Theory

Finally, because emotions relate to the interruption of goal states, I would like to mention how we can use this in defining customer journeys.

By way of example, lets consider my shopping trip to Morrisons, a UK supermarket.

Firstly there is the expectations moment of truth something that colours all my foregoing goals! This represents the emotional associations formed that colour our experience before we even think about going there!

Defining this opens up new goal states of emotional opportunity as Metro Bank found out in the United Kingdom influenced as it is by notions of greedy bankers.

Then once we have got beyond that, we need openly to pick up and group together in a nonlinear way a series of perspectives and narratives that represent our goals and sub-goals when engaging with the brand. For instance, forget about the journey; what do I remember about my last grocery shopping trip or I pick up 'in' the trip; what is important about these things; which are the most important elements and which are the modulators?

So, rather than construct things linearly – which requires a fill in the preformed gaps approach – now we set out a number of events, with visual description that defines our goal map by stakeholder. Some of these events are going to exist at a concrete level; others are more abstract; and all are going to be affected through time by what we remember (which is why firms should not ask for a survey just after purchase of my sofa but a year in as well).

Here then are some very brief emotional vignettes that represent goals and subgoals during my engagement with Morrisons:

I go to Morrisons because I love its convenience.
I always park in the same spot, so I remember where I parked my car. I don't get confused.
It's not especially exciting, a bit dull.
My kids like the magazine and toy area.
I use the self-serve as it's speedy, but not especially user friendly. I want to get out quick.
The quality of food is good; this is really important to me.
The staff wear those straw hats; it looks good.

Then there are examples of emotional omission

We don't go to the sweets area.
I like to look at the *New Scientist* magazine but no others.
I don't like to spend too long there – it's a hassle with so many people.

From these we can now look at creating a goal state map, and identify what goals we should amplify and dampen. For instance:

Things to Amplify

Core Goal: Convenience

Why can't we make it more convenient? I don't mean move the store, I mean in terms of perception. Clearly the car park is nondescript so putting some poles up with letters would help change perceptions of convenience, and make me feel more comfortable.

Brand Goal: 'Memorable Experience'

Why can't we amplify the 'experience'? At the moment it is a convenience-based value, but just like Luton Airport, if we could engage customers more by theming the sweets and toys area (unused) or provide cooking displays we might deliver more use more spend: the natural outcome of the value of time well spent.

Why can't we amplify the 'experience' of use-based segments such as the shoppers who repurpose the experience to browse (as in my behaviour at the magazine area)?

My engagement with the brand is also very habitual. Is it not possible to communicate through an app approach new items in stock, and I don't mean on my handset, but on the trolley itself. In the moment when you are engaged?

Things to Dampen

Core Goal: Reduce Crowding

There are a number of confusions in the purchase experience. The use of self-serve is one! There is a certain degree of corralling in one small location. Why? When there is space around the store.

Core Goal: Reduce Confusion

I always find it difficult to find certain items; is it not possible to download an app to say where some goods are?

There are in fact an innumerable number of things to change and adapt that could build a sense of place but must be tied together.

Think Holistically

Of course, Morrisons have actually tried to do customer experience, but what has been lacking recently is holistic design thinking. Companies engaging in emotional design must, like Disney, consider the whole experience and what they aim to achieve from it in terms of customer value; not a set of piecemeal activities. For instance, Morrisons undermined their CX programme with one particular piecemeal attempt: blowing mist over the vegetables.

Here is a comment on their mist spraying attempt:

> At the stores that had them, a number of people enjoyed the spectacle and took photos of the misty produce – but did it translate into a load of new customers? Not one bit. In fact, former chairman Sir Ken Morrison reckoned that it did quite the opposite and ended up alienating hardcore fans of Morrisons, who all buggered off to Lidl and Aldi.
>
> http://www.bitterwallet.com/morrisons-get-rid-of-dry-ice-veg-stands/83442

On its own it does do nothing but waste money: it has to be part of building a sense of engagement that leads to higher spend.

But I also strongly suspect there is an issue of gaming here! Leadership did not support it, rushed to ROI and made up their own story. There is no evidence of mist causing increased sales in Lidl and Aldi any more than there is evidence of mist on its own increasing sales in Morrisons. But plenty of evidence that such a piecemeal activity worked in part – customers shared it – only to fail due to leadership hostility to the idea of customer experience.

Disposition and Decision

Design then involves creating emotional actions that lead to dispositions and decisions. If we were to roll this up into a model we would find Fig. 10.2.

Here we can see how emotion's effect can be nonconscious or more memorable. This is shown on the x-axis. If we look on the y-axis we can then see its impact on ROI, which is either a direct or indirect effect.

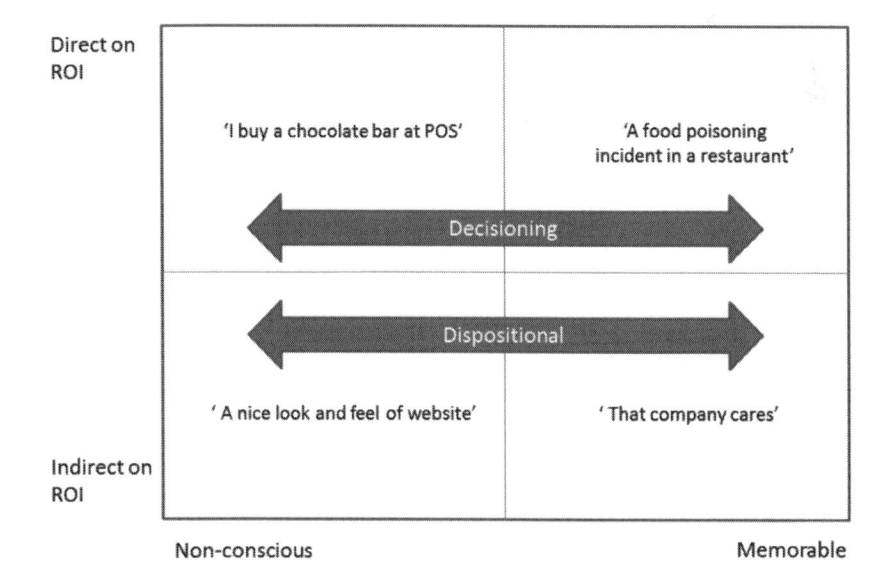

Fig. 10.2 Emotion framework

Hence, a nonconscious impulse with an indirect effect will lead us towards certain traits as we engage with an experience. Such as how we might notice the look and feel of a website. On its own, it may seem irrelevant but taken together these dispositions become more memorable. Hence, we see how moving to the right a sense of care is built up.

Again an indirect impact on ROI but nonetheless critical; there most certainly are correlates to spend here! No one wants to go to a dull store even if the price is right.

But we don't stop there.

The top two boxes also indicate how emotion at a nonconscious effect level can guide behaviour, such as I see a chocolate bar at point of sale, it arouses an interest and I buy.

Likewise, emotional effects can be memorably impactful on ROI, such as that restaurant disgusts me due to a food poisoning incident reported in the newspaper: I never go there again. Or we define a new emotional need! LUSH is emotionally engaging in a memorable way and we buy.

This brings me back to the point that emotions are personal; so start with what it means to be a certain customer personality or type and design from there. Or start with your purpose and theme and decide what emotional values you want to project. This is why you review voice of the customer data, undertake emotional journey mapping, and walk the experience: so you can be a better designer, not so you can measure, monitor, analyse and report!

Start with a deep understanding of what it is to be a customer type or how your brand needs to come across.

Uncover destructive and stressful experiences that degrade the emotional experience.

Uncover emotional drives partially established or needing amplification.

Uncover emotional drives not established and needing establishment.

Repurpose from other experiences, not so much to move the number but to move the experience.

So in short, it is not enough to talk about needs in a rational conscious way, as in I need a pair of shoes. Emotions can knock needs out. Likewise, an emotional awareness can move our dispositions towards a purchase! As the best salesperson will tell you.

Hence the need to be emotionally aware of the experience the customer has or could have.

But we must also remember that with CX as with loyalty it's not about cognition **or** emotions; both are equally important as is the behavioural dimension, what we do. (Olaf Hermans)

Bio-Connectedness

But there is one thing missing in my discussion of emotion. Because emotion is a prototypical word, we should also think of biological data, which leads to the question, how does this sort of more objective information about our self and how we feel relate to action? For although there are challenges in terms of self-report, other related aspects of emotion can be more easily accessed; for instance facial coding although limited gives indicators of nonconscious emotion states while watching adverts, likewise FMRI indicates feelings related to brand experience beyond the reported concerns on price and functionality; EEG on stress.

So what else is on the agenda?

What else connects our biological and emotional reactions to data.

Well, hospitals already connect up our vital signs to data and stress and heart rate monitors are already out there. So what if this was made easier? The data stream is now, perhaps not how we feel, but certainly indicators of how we feel. So picking up spikes in our stress rates fed to my own personal health monitor I might relate this to life events and in a self-serve way ask for assistance (once I have had a chance to reflect on the data!).

We have seen this possibility with up-dosing and down-dosing drug regimes, but this all applies in contexts such as early warnings on high stress rates (which I can report through say an app) and how this is affecting overall health (which maybe I can't) requiring lifestyle alterations.

No one is suggesting firms gain access to who you are, at least in a quantitative-self way, but as with any data layer the consideration must be, 'Can this be turned into value?'

Management Implications

1. Emotions it seems, 'are not a luxury, they are essential to rational thinking and to normal social behaviour.'
2. Traditional frameworks and survey approaches assume emotion is something objective not subjective! This is fundamentally wrong because it ignores the semantic content of an emotion and its personal subjective nature.
3. Emotions are a cognitive agitation of the nervous system which means they are *fleeting* by nature.
4. *Fleetingness* is a surface feature of our relationship to a brand, which again causes problems for research.
5. *Fleetingness* can also mean emotional effects are nonconscious, which means traditional research simply fails to report them.
6. *Fleetingness* causes us to under-report, over-interpret and nonreport emotion! It also means we cannot reconstruct emotional feelings by getting customers to respond to some survey served up three weeks after the decision event.
7. Just putting emotion words on a scale and asking consumers to rate them is not the same as meaning they feel them![2]
8. In research, use alternative approaches that are designed for fleetingness. For instance, immersion, taking narratives from multiple perspectives – consumer, employee and experts – as they walk through the experience; more realtime and immediate gut reaction/nonconscious measures.
9. Emotions act as 'weak signals', they let us know what the 'experience' means and where it is heading, in other words, its disposition.
10. Dispositions can also be affected by the mood of the environment, which firms do have some control over.
11. If we could measure 'resilience' we might find that what seems like a strong so-called promoter market is actually not so emotionally engaged.
12. We can also consider using value-in-use frameworks to unpack hidden dispositions around anticipated emotional reward.
13. We must go beyond traditional research approaches which only use complaints data, social media and surveys to measure emotion while comprehensively missing its *fleeting*, hidden and dispositional nature.

10 The Value of Emotions

14. In the immediacy of a situation, emotions certainly have direct root-cause effects. Emotions also indirectly affect value.
15. The real value of emotions comes from how we can use them to identify a new emotional value.
16. Develop a deep intuitive connection to the customer.
17. We are not looking for an emotion; we are looking for more meaningful experiences that relate to what drives us as consumers, from which an emotion is derived.
18. I would argue we should use the term *drives* not emotion.
19. Emotions, to state the obvious, are personal not controllable.
20. Avoid loss aversion. Loss aversion is also about '*proactive*' loss aversion.
21. Be personal.
22. Be sociable.
23. There is one area that we do control, that is the physical or digital environment through which a consumer/employee engages.
24. Respond to dispositional characteristics of emotion. This is being aware and responding to how things change in the emotional environment, understanding what to amplify and what to remove, not waiting for something to happen; identify it early.
25. To understand nonconscious effects it is useful to immerse ourselves in 'what happens' and observe customer behaviour in the experience.
26. We should also create memorable effects; things that 'stick in the memory' and reinforce a good impression of the brand.
27. Manage stress.
28. Alert and understand specific emotions. 'Start with what customers want to feel or think they are going to end up feeling – and don't treat emotions just as a "positive" vs "negative" that is archaic – if your angry or frustrated you are still connected – you still want to tell that brand – if you're bored or indifferent on the other hand then the brand has just lost you….'
29. Measure emotions effectively.
30. Account for heuristics and bias.
31. Uncover what drives the customer. These are based on goal states. Goal-state map!

32. Emotions are personal, so start with what it means to be a certain customer personality or type and design from there.
33. Think about your brand theme and purpose and what emotional values need to be consistently designed into your experience.
34. 'But we must also remember that with CX as with loyalty it's not about cognition **or** emotions, both are equally important as is the behavioural dimension, what we do' (Olaf Hermans).
35. What else can bio-connectedness do?

Notes

1. There is evidence that people spontaneously regulate their emotions (Forgas and Ciarrochi, 2002). Immediately after an emotional event, people in both happy and sad moods experience more mood-congruent than mood-incongruent thoughts. With time, however, the content of people's thoughts moves towards the opposite valence. That is, after a few minutes, participants induced to feel sad were having happy thoughts, whereas those put into a happy mood had relatively more sad thoughts. This homeostatic emotion regulation fits nicely with the current analysis: mood-congruent thoughts help people learn the lessons of their previous behaviour, but adaptive future behaviour requires that emotion regulation take place.
2. The irony here is that NPS does not measure dissatisfaction and a bipolar CSAT scale means consumers answer towards the neutral because a call centre has aspects of satisfaction and dissatisfaction, or answers tend towards the positive because you ask me about satisfaction! And I gift you an answer.

Part V

Mindset

In this section, we turn our attention to the internal organisation. Employee experience and culture are critical to delivering the Experience brand. If you don't achieve the right mindset, all your customer experience actions will be for nothing.

11

Right Mindset

Mindset, that's the word for it. How what you do for a living dictates how you see the world. It's not about how you are incentivised.

So a chief technology officer's mindset will be focused on the hard concrete reality of numbers and the look and feel of say a computer screen or a server.

The beauty they see in these features is after all why they have succeeded.

And because this approach is the path to progression, the CTOs surrounding business culture will be infused with such thinking. The language they use both formally and informally, the power statements and arguments as well as their prejudices will be constantly reinforced by this way of 'paying the mortgage.'

Hence, in this paradigm the subjective experience of customers is not important – that is just fluffy data – and instead CX is all about the quantity of things, risk assurance and machine data. I have even seen organisations such as the TM Forum engage the term customer experience while completely excluding the customer! Assuming the functional and the hygienic equates to relationship value!

At least I ask myself, why bother? Just call it efficiency.

A CTO then is most likely to say, 'Customer experience, that's an IT box,' something that measures and sums all touchpoints.

Here, you may find £1 million spent on a system of measurement that is really all about the CTO and his or her risk averseness and love of visual IT architecture diagrams; who said CTOs were unemotional!

Similarly, for a CFO mindset means cost-benefit calculation, focused on financial root cause.

A CFO or CEO is most likely to say, 'What's the return on customer experience?' 'How much will it cost?' and above all else, 'How can we save money by cutting costs?' For them, intangibles are woolly and fuzzy. Hence, culture and subjective experience are not of concern; and so too, by consequence, relationship and loyalty, the essence of CX!

For CMOs or sales directors the mindset means how much they can sell to the customer.

So:

The CTO is out to spend as much money on IT as possible.

The CFO is out to save as much money as possible.

The CMO and sales director are out to get as much money as possible from consumers.

But where in all this is the customer experience? If it's all about the experience the customer has, no one seems to care!

We could say that the insights director is important: speaking the voice of the customer to the organisation. But for the most part, 'insights' in business provide sales messages.

We could say that the operations or service director should take the role. And to a certain extent they do! But this usually revolves around controlling what is measurable, that is to say complaints. Their voice is constrained.

We can see that for each of the powerhouses: CMO, sales director, CTO, CFO and CEO the mantra is 'What gets measured gets managed.' No fluff; it must be tangible. The problem is this mindset assumes that in customer experience only those things that are objectively measurable such as price sensitivity and efficiency gains are important, and that what is measured is unaffected by the less measurable and intangible 'soft' stuff!

Or at least measurable but in a different way!

And this creates several problems.

Firstly, just because something is objectively measurable doesn't make it valuable. As we have seen with the root-cause focus on loss aversion, we constrain ourselves if we get fixated on only the measurable thing. After all if you miss out the emotion/drives of customers you fail to understand how value is defined and decisions made with the intangible in mind.

Secondly, if a leader wants a *measurable* number increased, you can guarantee it will happen. Not because of some improvement in activity but because the number has been gamed. A higher NPS score leads to the achievement of a bonus. And I get my promotion. Here an NPS increase is assumed always to be a relationship currency rather than something gifted, fleeting and forgotten.

Hence, *measurement isn't management*; intangibles such as culture and customer perceptions are critical. So just because their measurement is difficult does not mean they are valueless. Unless you want to believe that only measurable quantities matter, which sounds like madness. And it is not just due to complexity; it is also due to the simple failure to record losses: the losses due to a bad boss who causes productivity declines, increased employee turnover, a yes-person culture and a fear of innovation; the loss of sales due to pushing product rather than solution selling and the failure to engage customers due to a focus only on reducing complaints.

All these and more are never accounted for unless we consider the intangible.

And anyway, the full proposition is: 'What gets measured gets managed – even when it's pointless to measure and manage it, and even if it harms the purpose of the organisation to do so.' As Henry Mintzberg says it would be better for us to start with the belief that we can't measure what matters; that leaves us better able to face the challenges of managing a business. A measurement you see, can be a qualitative observation, an expert opinion.

If Opex reduction, sales growth and process efficiency are your thing who am I to say any different. All I am saying is with a clear view of what customer experience is we may be able to gain advantage: so understand it not over-brand it.

I for one would be quite happy if customer experience became a lot smaller but better defined.

So for the company that wants to be an Experience brand there are many challenges, but there are many examples of companies that have chosen this route and been successful, companies that have changed the mindset of the organisation.

And what do these companies do?

Start with Understanding

They start with an understanding of why they are doing customer experience. Remember you don't have to do it! But if you do, be serious about it. For instance, one pharmaceutical company I worked with spent upwards of £1 million on a yearlong project looking at customer experience. By the end of the project the team had been disbanded and the project shelved. You might as well have gone outside and burnt the money.

By contrast Metro Bank saw the emotional value of trust as important in an environment where banks had failed and they deliver against it. Go into any branch and you can see they have taken the experience route.

Obtain Budget for Actions

One of the constraining factors I have noticed in customer experience programmes is how everything tends to be built into an industrialised process, the heart of which is usually measurement. But I believe greater balance is required here, between expenditure on measures and expenditure on actions.

Change the Corporate Mindset

Side by side with this, we must also engage in strong efforts to change the corporate mindset.

And particularly with legacy organisations this must start with the leader's actions and behaviours. They must show that customers matter.

Without this a new mindset will never gain traction, however much the mid-tier believes in it. And they must walk the talk, communicating their belief in experience throughout the organisation.

Leadership of customer experience also needs to be co-operative. It has to be because it is about cross-silo and cross-business cocreation. Eighty percent of the leaders' activity needs to be on the ground dealing with human issues; 20% needs to be about the task itself.

For instance, bad leadership led a large chemical firm to tell its lab workers to build a new compound. The workers themselves knew this was impossible. In spite of regular reports that this was the case, no remedial action was taken. Results were gamed; leadership was congratulated. Politics was rife. This is not a customer experience approach.

Embed Process and Governance Change

Customer experience ways of working must also be allowed to seep into how processes are designed and how teams are built up to operate within a light hierarchy with measures that are loose rather than tight: overmetricising creates a restrictive, myopic bonus culture.

All difficult things to encourage when faced with the inertia of a hierarchical, process-orientated company set up to manage the demands of unitising, scaling and pushing the sale. If Experience brand status is about establishing a relationship with your customers and employees, then things will have to change.

But I don't want this to sound like we are heading for a nirvana of happiness! Some things are impossible, but I would argue even within legacy structures there are also possibilities.

So maybe it might not be feasible to reorientate your account structure away from individual silo behaviour and more towards group incentivisation based on customer perceptions of the whole journey 'drives' (e.g., goal states) (or) reorientate your firm towards customer benefits sought rather than product silos! But that doesn't mean you can't put into place aspects of CX governance that do affect the direction of corporate travel!

Create a Small, Passionate, Cross-Functional CX Programme Team

With customer experience programme teams, small is beautiful, because small teams offer the highest degree of agile learning, flexibility and the capability to build cross-functional links so essential to customer journey redesign, as long as team members have influence and budget within the organisation and are interested!

After all we are not creating a big business here. In RBS the CE team comprises about 20 troubleshooters[1]; in others the team is even smaller. ODC Bank in Singapore originally had a small team of 15 highly focused on redesigning journey pain points. And these are big corporations of hundreds of thousands of employees.

Small teams also have the benefit of fast piloting and continuous feedback. For instance, this is how Lockheed Martin are trying to build a fusion engine. They realised that large machinery to produce energy took years to build, years to maintain and limited the flexibility to learn, adapt and be creative. This last point being particularly important if you think how, because of the time, effort and cost built into running a large reactor, every change has to be carefully planned and every change takes huge effort, by which time management has moved on.

Hence, customer experience ways of working are far from being abstract. But it is also important to realise that customer experience isn't a job creation scheme. By talking about light 'narrative-based' KPIs, sometimes but not always limited investment in IT infrastructure, flexibility to pilot 'actions before measures' and small communicative teams, I am talking about customer experience as a small discipline that drives design and mindset changes in the firm.

Be Customer Advocates

With customer experience then, we 'promote the customer first view and create interconnections between departments around this view, acting cross-silo (and cross-company) as a troubleshooting and piloting team, maintaining the customer and employee story.

Silos after all can be great; it's the *interconnections* we need to create through interaction and data that are important. I am not one of those who believe silos need to be broken and everyone in the IT back office should think customer. This for me is unrealistic, constraining and can lead to perverse consequences. Now every breakage, every detractor is seen as a terrible calamity; my God that's a detractor! Well perhaps they deserve to be, and good riddance. We need appropriate de-siloed thinking; appropriate levels of interaction across the business and appropriate levels of think customer! Not a wild-eyed dictat.

If you force CX too much on people you will get a negative reaction, undermine and degrade your case.

CX then is the one part of the organisation that takes the customers' and employees' side, to balance the risk of companies myopically focusing on themselves, the business account and closing down creative potential. We are the customers' advocate at C-suite level and through champions within the firm. And I suggest at the interface with the C-suite we don't just recruit people from within our industry; we need to challenge thinking not just do the same old same old.

Market the Experience

It is also a discipline that needs to be in marketing.

It has been said that 10% of a marketing budget should be set aside for 'the experience'; this is something I agree with and maybe if you don't agree with setting up a separate design team, a useful place to start. For instance, some operators enable customers to go into retail stores to try out their new phones in Experience Hubs. Likewise many B2B companies provide Experience Centres, where clients can come and 'get an experience' of using the products. An example of this is Jaguar Land Rover.

Be on Brand

Finally we need to focus what customer experience means to the brand. Ultimately you need to answer the questions: 'What are the Experience brand values you want people to talk about? How is this reflected in behaviours?'

Here are a few examples then of how mindset change has been successfully initiated within corporate organisations and supported the Experience brand:

Mindset Change: Mean It

Many firms have ill-defined their purpose, usually mentioning terms like we want to be loved or most trusted. The problem with this is that without clear directionality, how can your brand execute the right behaviours?

For instance, I worked with one Indian firm that clearly stated their intent: to be the most loved brand. Now with words like that, you have to question whether this is really all lip-serving; but never mind at least they were thinking customer. Except that when it came down to brass tacks, 'love' was not an invested behaviour. There was no response to data that demonstrated a failure at the customer service and billing end and there was no interest in connecting network engineers to the CX stakeholders.

Clearly, 'most loved' meant nothing at all!

And I think the reason for this had been again the rush to purpose. There had been no unfreezing of the organisation and the inculcation of a platform for cultural change and learning prior to throwing marketing concepts out there and hoping for the best. This for me is a fatal mistake: you need to start inside the organisation before you can hope to do CX.

Mindset Change: Be Consistent While Maintaining Flexibility

McDonald's and Disney are brands with clear directionality and consistency of approach. They understand that it's what you do that really matters, the behaviours not the metric. Contrast this with brands that are either too strict in defining their criteria – hence failing to evolve and see how markets change – or lack consistency to define and drive through a clear CX approach. Finding this balance is crucial. Who wants to work in a firm which gives you no room to breathe and likewise how many customers want to pay for a brand experience in one store only to find it's different in another.

In one Thai-based operator, for instance, I found their CX programmes completely disjointed and all over the place, a bit of support here, a bit of money there, a head of CX with no power but under the remit of the call centre manager. This does not help spread belief in the programme. Likewise with a Dutch telco, a UK financial services provider and a global transportation company, a bits and pieces CX approach was enacted. In this environment, without leadership traction, any efforts will die from disillusionment.

Mindset Change: Focus on Customer Value (Move the Curve Not the Metric)

What is important is a mindset focused on investing in improving the customer value (to cost) equation. Bearing in mind that much of what is 'of value' is ill-reflected in traditional driver analytics which tends to ill-consider fleeting affects (modulators, emotion, the nonconscious); and just as important fails in its focus on 'as is' drivers to consider that these may lack differentiation (all operators offer good network reliability) and that the same score could drive different value as they mean something different, as in Harrods' food quality is 8 out of 10 as it is in Sainsbury's but each number means something different.

Mindset Change: A Catalyst for Changing the Contract Relationship

Run by a small passionate team, a leading, global B2B logistics player operates a CX programme focused solely on helping clients deliver a better customer experience. Through this programme, they aim to build a closer more strategic contractual relationship that goes beyond the delivery of 'business as usual.' As they say, one that innovates new solutions and 'hunts the farm.'

Of course in this way the company hopes to find upsell opportunities and improve the level of contract retention and satisfaction. But the point is, this is not a sales plan: they are not pushing out multiple variations of products and services in the hope of getting a sale. Instead the firm operates outside of sales through the service function in order to enable time, consultation and cocreation.

And this has been a major success. Solutions have been designed that are light touch, such as how they helped one client gameify their website, or much more involved. For instance, with one client they signed up to their NPS KPIs even though they knew a lot of influences on NPS were outside their control. The latter of course was a managed risk but the point was that if the client could not pay the bonus that opened up the conversation to why, and how they could help them further.

In addition, the business has from day 1 given clear sway to the business line to prove its case, ring-fencing a set of key accounts over a two-year period to test whether improvements are apparent and seep through into ROI.

The CX contractual process is also very visible with materials specifically designed to reflect the different nature of the relationship. It is also supported by regional customer experience champions, an intranet training site and clear leadership support; for instance customer experience is set as item number 2 on the board agenda after finance.

Of course with a legacy environment, a major focus for the team has also been to spend a large amount of time educating the business on the programme and spread the word about customer experience.

The programme therefore shows a key point about customer experience. It is a relationship sales process. So innovate–collaborate, ask the question 'How can we help you?' the customer not 'How can we help ourselves?' Hence, define the customers; engage with them to understand their drives and deliver solutions to help them.

The need is to mass customise not mass produce, to quote Joe Pine.

Mindset Change: A Catalyst for Changing the Way of Doing Business

Here Rob Frank, customer experience director of a large UK infrastructure and construction company, talks about how they changed the way of doing business through their Perfect Delivery value proposition.

'Client relationships have evolved several key differentiating service excellence principles:'

1. Traditionally construction is a low margin business with returns gained from an on-site confrontational environment. We changed the experience through service excellence, offering delivery "on time and to quality." This was most definitely personal and memorable and they were able to focus here because they did not have a legacy mindset (something Rob and his team relentlessly work on). From day 1 they emphasised their core principle of 'on time, to quality' delivery (Perfect Delivery).
2. Over time this morphed into a concern for how they could help clients in their customer experience: something called the 5th Pillar. This doesn't have to be a huge exercise; it could for instance be as simple as better snagging at the end of projects.
3. Beyond the contract relationship the company moved to look at how they could add value in terms of the practicalities of doing business. This led them to maintain a database of client interests; for instance if a client held an interest in sport, they offered tickets to key events. It also meant taking their voice of the customer reports seriously.
4. The practicalities of onsite business were also altered to reflect the customer experience focus. And this didn't just mean a mindset focused on doing a good job; it also meant delivering service excellence to all parts of the onsite relationship. Hence, the company now focuses on the experience of employees and contractors. Go to one of Rob's sites and you will find an onsite concierge-like service for hardhats and jackets, great facilities as well as the best sandwiches onsite! The aim is that onsite suppliers will start to talk about their great experience, and as we have seen, suggest them for further work as they work between clients.

'We focused on the employees experience; the project managers, quantity surveyors and mechanical and electrical engineers. Here value was added by creating a great on-site experience: putting in place 'concierge' services, having great sandwich lunches, ensuring the facilities for the employees were over and above expectations and so forth. The idea was that since these intermediaries enjoyed working with the company they would be more likely to suggest them as a supplier. In fact it was later worked out that 10x the value was generated by these intermediaries over end-clients due to their influence on future tender selection.'

5. Rob introduced a small centralised team that spends a large part of their time communicating with sites about what customer experience is and how to engage with their B2B clients.

Hence, customer experience has two foci, one on the contract relationship – Perfect Delivery – and one on the experience of doing business, that is, at a supplier and site level as well as between the company and its end-clients.

The service excellence mindset didn't need an ROI justification before proceeding; the customer was inside as a strategic imperative from day 1.

So now, although of course they refurbish offices, they also sell employee comfort. Employees like the firm as they provide the best site to work on: nonconfrontational, great sandwiches, concierge services, a pleasure!

The service excellence mindset didn't need an ROI justification before proceeding; the customer was inside as a strategic imperative from day 1.

Mindset Change: A Low Focus on Heavy Metrics

Which is better, to have two people in a business enthusiastic, empowered, with a budget and with a clear sense of direction based on 'right and light' measures, a design focus and the capability to evangelise the message or 200 people measuring and monitoring activity?

The best customer experience companies of course put voice of the customer information front and centre stage. But they also understand that customer experience is not a machine-like system where you pull one lever and out comes a higher NPS score. They understand the need to engage and flex, engage analytics effectively, use quality data not reams of data and perform actions.

Hence, Capital One and JCB deliberately enable time for their executives to listen in to call recordings. Build-A-Bear and DHL use the customer information to engage and collaborate with their clients, suppliers and customers to create solutions.

The best firms also understand that the best data involve sharing customer stories, that there is a need to get away from statistical abstractions and really experience what it is like to be a customer. Hence, they immerse themselves in the life of a customer, understanding how their products and services are used.

A Swiss insurance company, for instance, regularly uses an app to get its employees to engage in small customer experience tasks. They want their employees to know first-hand what it is like to be a customer. Understanding that immersing employee 'experts' can change the company's understanding of the customer experience and deliver new solutions which customers may not have thought of.

This is something I call developing the creative equity of the firm (as opposed to the usual business of expending time and energy on analytical equity).

Mindset Change: Have a Good HR Team

An employee environment is required that can manage flexibility, adaptability and change. Hence all that we have said about customer experience techniques such as mapping and measurement applies to the employee environment. But most critical of all, is probably having a good HR team, one that seeks to build the employee experience rather than is focused on transactional functionality. If there is one thing that a company can do in CX, above everything else, it would be sorting out how employees are engaged and inspired.

For instance, in the rush to Opex reduction, how many firms force self-serve on employees for every action. Or destroy the informal network by over use of virtual systems. Or take out the supportive function of HR?

In one company I worked in, HR from start to finish was nothing more than an administrative function. On the balance sheet, costs savings were being made, but none of these costs savings accounted for the loss in employee time and the degradation in morale. Even worse, any failure of technical systems was actively pushed onto the employee.

These kind of behaviours are symptomatic of failed leadership. One that is incapable of seeing the critical value of culture.

In part, this is an endemic reflection of over payment, which causes some leadership to believe they are the only source of value. They look at certain geniuses and believe bad behaviour is what delivers results. Not realising that they are the exception to the rule. They get away with it because of their capability in other areas.

Hence the need for a strong HR function, to counter-balance the worse excesses of leadership and provide a roadmap for the employee experience. And hence the need to drive home the message that 'what gets measured, is a poor reflection of what needs to get managed', that measurement is also an observation and management a facilitation: so if you as a manager can't observe and can't facilitate then you should not be in leadership.

Mindset Change: Bring in a Third Eye

If the business is one eye and the consumer is another, it is essential to bring in a third eye outside your industry. This could be through looking at best practice customer experience from other industries or engaging with experts such as anthropologists or behavioural psychologists.

This has the benefit of shaking up the mindset and finding answers to problems not thought of before.

Mindset Change: Build an Evangelist Network

Formalise and train a CE network of evangelists within your company. This enables a community of interested experts to develop and gain momentum. It also helps develop practice. For instance, learning how to benchmark best practice customer experience or understanding agile, service blueprint and complexity methods, a network to share stories and link up with key academic institutes in training and learning.

Mindset Change: Promote Cocreation and Go Beyond the Silo

If you are collecting insight, you may find that what customers and employees want goes beyond what you currently offer!

For instance, in looking at a network experience for an IT provider I found that network perceptions were driven not just by the network features of integrity, reliability and accessibility – that's engineering talk – instead customers frequently spoke of the need for better marketing and customer service.

To execute such insight requires collaboration because you are going beyond a feature and function sale where cross-silo working is key. You may think in silos, customers don't!

Mindset Change: Relationship Sales, Not Sales

In customer experience you leverage the relationship to uncover deeper drives. How we can help you in your experience; how we can engage customers at a more motivating level?

This therefore necessitates a relationship or solutions sales approach, being able to engage and talk about clients and customers' business problems not just push a product. It also means showing some challenging thought leadership to be seen as a trusted business partner.

So one pharma company in moving from selling blockbuster drugs to selling the relationship used a solution selling approach to uncover services they could provide that helped their clients, and one that also demonstrated their credibility to move beyond selling pills. Hence they have been able to sell smoking cessation programmes to doctors' surgeries, something that is meaningful to this client base.

Mindset Change: Develop Your Creative Equity

Companies are quick to measure what is, but fail comprehensively to use the incredible creative resource of customers, employees and suppliers. In other words they are high on analytical equity, low on creative equity, which is a shame, considering the mine of information you have out there.

So, don't just do text analytics and surveys, create community hubs, generate ideas through engagement and build in Experience Design programmes; programmes which consistently fast-prototype ideas through to trial and test, understanding that fast failure is the way to develop success. This after all is how LVE, the insurer, won the UK CEM Awards with their !NNOVATE programme for employees, and how Starbucks use social media to help in product and service design, such as how they created the Via coffee sachets and new varieties of coffee.

Creative equity is a vast untapped resource, and not just from your own network but through combining different industries together in creative ways, such as we saw with the use in construction of the Mandarin Oriental concierge idea.

Mindset Change: IT Is Not the Answer, Alone!

IT automates; that's what it's there for! It is not a substitute for human understanding. So don't just automate your journey maps; keep them graphic and engaging: use video for instance, anything that will drive learning amongst employees. And don't just automate in a big data solution; use the employees' understanding of the experience the customer has as well; better data are more important than big data.

IT without humanity depersonalises, which is why it is critical to have the customer experience eye on any implementation. For instance, look at the depersonalised job market. The Internet has created a faceless agency behind every job ad; why? They don't add any value; they have no idea about the job or the culture. In fact they create a barrier, forcing candidates to go to interviews in part due to the fear that 'If we didn't go' these people might not put us forward for other roles. Or putting people forward who are wrong for the job because the agency has no idea about the job. Or being too restrictive in their criteria, not seeing individual potential over prior skill.

Although of course, agencies that are good understand this!

Or how about the way automation has led to the commoditisation of the employee experience: put stuff on our intranet site; make it 100 pages long when 2 pages would do. Send out automated responses to queries.

Make any administrative activity a maze of IT systems, not because it benefits the employee but because it saves money; except it doesn't! You are not accounting for time spent working our systems and a poorer work environment.

Remember it's easy with IT to add in functionality without consideration of the effect.

Management Implications

1. Mindset, that's the word for it. How what you do for a living dictates how you see the world. It's not about how you are incentivised.
2. Measurement isn't management.
3. Start with understanding.
4. Obtain budget for actions.
5. Change the corporate mindset. And particularly with legacy organisations this must start with the leader's actions and behaviours.
6. Customer experience ways of working must also be allowed to seep into how processes are designed.
7. Create a small, passionate, cross-functional CX programme team.
8. Mass customise! Not just mass produce.
9. With customer experience promote the customer first view and create interconnections between departments around this view. Acting cross-silo (and cross-company) as a troubleshooting and piloting team, maintaining the customer and employee story.
10. CX is also a discipline that needs to be in marketing.
11. We need to focus on what customer experience means to the brand.

Establishing a good relationship between employees and leadership through the right cultural framework and supportive business practices is the imperative. Easy words I know, but I would argue relationship skill is of equal importance to accounting. The serendipity of relationship is a sorely underestimated source of value.

Great CX companies invest in artisanship; they understand employees as a source of value and innovation not as a cost. After all, without engagement and some passion for what you do, you will just treat the job as a function and not invest in the behaviours required for CX.

Notes

1. Indeed, one of the claimed best examples of customer experience comes from the Royal Bank of Scotland. Here they brought in an engineer to manage the programme who focused wholly on cost reduction, Lean and Six Sigma. The objective became not 'What can we do for the customer to increase the subjective asset, customer experience,' but 'How can we benefit ourselves, cutting out processes that seem superfluous.'

   The case study was a great success but was it from a customer experience point of view? Yes, it reduced cost but my question is: Did it deliver an appreciable return on subjective experience? Furthermore, who is to say that some of these invisible cuts might well have been important components of a customer's 'experience,' or set the platform for more experience creation? I would contest that without a valid multimethod measure of experience no one could tell. But its use of activity-based costing is a significant addition to the CEM literature.

Part VI

Not Do

In this section, we look at some examples of bad customer experience that come from the firm efficiency-only approach. If this is what you are doing, then you need to rethink why you are even bothering to waste your time on customer experience!

12

Customer Experience Bad

So now that we have an impression of what customer experience is, why is customer experience so bad? And particularly from those companies that claim to be doing it!

Well my belief is that this is because customer experience is being defined through the lens of firm efficiency alone. And this means that experience is not being considered as a subjective response, but more as a sum of everything.

But how can this be? Shouldn't it be clear that customers do not measure every touchpoint, frequently don't care about efficiency and really only start noticing 'the experience' as something of value to them once you start to engage and motivate them?

Well, clearly not because this efficiency-only definition remains alive and kicking, churned out by innumerable vendors and consultants.

If you don't believe me, have a search on Google under the term 'customer experience'. You will find this turns up a vast array of software and technology vendors all promising the same thing: 'experience is everything'; just buy our software and we can measure everything. Indeed, much of the technology that had been branded as CRM, marketing analytics, campaign management, contact centre software, voice of the customer, sentiment analytics, churn analytics or whatever is now CEM. I am not saying all this is bad CX, some is perfect for CX, especially where loyalty marketing and personalised campaigns deliver a personal and memorable result! But the market is frothing over with solutions; we need to sort the wheat from the chaff.

The reality of this approach was further driven home to me when I met the director of customer experience at a world-renowned VoC software company. Customer experience for him was very clearly about everything, and because 'experience means everything' we could of course put in place measurement systems to, well, 'measure everything!' And build consulting teams to sell the software.

Drawing this new paradigm of business on a piece of paper, he started off with how experience was currently being measured and quickly turned this into a flowchart of boxes everywhere. Each one an extra sale!

But beyond that, the use of the term is also gaining great traction in other roles.

Look at any job site and you will find thousands of roles that have been rebranded in this way. You're no longer in customer service or marketing; you're in customer experience. Order chairs for a new mall: that's customer experience; answer queries at a call centre: that's customer experience.

Going into my local Royal Mail depot to pick up a parcel, a plaque was on the wall to say how staff had been through 'customer experience training,' no doubt from the Institute of Customer Experience, previously known as the 'Institute of Customer Service.' I can assure you the training is exactly the same!

Likewise, metrics systems such as net promoter and customer effort score now proudly proclaim themselves as customer experience metrics.

Many leading firms have also jumped on the bandwagon and rolled out customer experience programmes. These all follow the 'experience is everything' definition.

Here are a few examples.

Telecommunication providers as you would expect, take a very objective and tangible view of the world. If it can't be touched, it isn't of value. It's a point of view driven by the mathematical mindset of CTOs and solution architects, used to building networks, IT systems and computers. The language of correlation, process, cause and effect and predictability is embedded in the culture; the people who live in its world take pride, almost arrogance in their worship of 'the tangible.'

Perhaps the ablest demonstration of this mindset is to be seen at the Mobile World Congress held each year in Barcelona. Here CTOs (chief technology officers) from around the world gather to view the latest and greatest technologies and hobnob with the best talent in the industry. They are not here to listen to concepts or see displays on intangibles such as culture; they want to see a machine 'blinking' in the background with pretty lights and a demonstrable – and partly gamed – ROI. 'Measured and managed,' that's their mantra.

Here customer experience is a piece of machine data that measures and monitors the customer experience; it is not the customer that measures the customer experience! And if the customer is considered at all it is as a minor additional metric; a CX professional here is most likely to say, 'Stick some NPS measure in, that'll do! Our real focus is on machine data.' When a threshold degrades, that's an experience; that's all that matters. You can always tell if someone is concerned with CX or not: they will start with the customer and use technology as an enabler, rather than the other way round.

By contrast I contest that the experience the customer has is not the 0s and 1s of a data warehouse; it is far more qualitative than that.

But don't get me wrong: machine data are good; I like the innovation push from continual attempts to commoditise and the focus on controlling loss aversion. But it also blinds us to the experience the customer has because without the integration of perception metrics how can we know whether what we see as an alert to a threshold is important, or whether an efficiency drive is leading to a positive or negative customer reaction? Or whether we are wasting huge sums of money on hygiene for negative return!

Also, where do we stop if we don't include the customer? I mean a bill reaches a customer, so is the printing machine for bills a CEM solution? Is the ink? And if we don't include the customer our definition of the objective everything misses out intangible value. Perhaps the way the bill was printed, the way it was written in a formal style is the problem. Not the fact that the phone works.

Without this understanding I have even spoken to operators' marketing departments where they claim getting a signal as a moment of delight!

We also miss out on the new things we could do to engage the customer. And as with CTOs so with accountants.

I went through the interview process for a large insurance company based in the United Kingdom. This was for the post of director of customer experience, in reality a mid-tier job of little consequence, because like most firms the company just dabbled in CE, doing a bit of journey mapping here and of course obsessed with NPS. In fact the job could have been described as ticking the box by buying and integrating an NPS software system.

The meeting was once again confirmatory; customer experience is about measurement. Maybe these people are not so enamoured by a piece of technology like in CTO land, but they are as accountants fixated by numbers, cost reduction and 'gamed' NPS scores.

A typical CEM programme here involves delivering NPS scores to each and every call centre agent with associated verbatim, a programme essentially of efficiency and reducing any points of conflict rather than adding value to the customer's engagement. This is employee engagement by numbers not by inspiration, empowerment or artisanship and one that only leads to siloed behaviour: I'm concerned about my NPS number not yours. As well as restricting behaviour: I must treat each customer in a fixed way.

So where in all this are you 'throwing the Pike Place fish'?

Think about it this way: if we go back to Pike Place and now apply this efficiency only, or as we saw 'little e' approach, what would we have found?

Certainly there would be no focus on attracting customers through memorable moments. Instead there would have been a large-scale investment in NPS (or other) software to measure every touchpoint with fishmongers incentivised on their ability to get customers to score 9s or 10s.

Of course, journey maps would have been invested in, bosses would have got their bonuses but taking cost out would have been the name of the game not adding value. The emphasis would have been on greater firm efficiency, as if measuring a boring experience is actually going to leap you into profit. Or a piece of machine data with attached PowerPoint slide pack that promises some algorithmic relationship to customer perception improvements is actually equivalent to what really goes on.

So any frills and fancies around making the experience entertaining would have been removed. Pike Place would have spent money on systems that would have pushed out junk mail to the inboxes of customers.

Without a shadow of a doubt, the best fishmongers would have left and there would have been no focus on 'throw the fish'; hell, what's the ROI of that! Pike Place would have gone bust.

So what can we conclude?

Experience is not about everything because not everything matters. When I walk into a fish market I don't walk around scoring all 'experiences' out of 10. When I walk into a fish market I don't ignore all human interaction and emotion and only focus on the physical touchpoints. When I walk into a fish market, I behave as I have always behaved: some things count, some things don't. Mostly my experience has nothing to do with technology and sometimes technology is an enabler of a better experience.

Customer Experience Bad Approaches

Defining CX as: Process Only Firm Efficiency

This is an attractive definition of experience for many supposed CX firms because process is scalable, tangible and focuses on hard metrics such as Opex reduction: all the things of benefit to the firm's bottom line but risk degrading the experience the customer has!

Here are a few examples:

- A well-known bank closed their branches and moved their customers onto online channels. The Opex reduction, the ability to control waste (because digital is measurable) was clear and attractive to the firm but the customer view was ill considered. For instance, would the online channel be easy to use, would the bank still be able to answer all the questions customers ask on the phone, how would the changeover be controlled and what if customers wanted to talk to someone? Overall in its delivery the short-term cost advantages blindsided executives. The irony being that managing these new channels to the level of the

branch experience will be just as costly especially because these channels will be scaled, commoditised and hence lose their competitive advantage!
- A telecommunications operator invested in social media forums with the intention of reducing the amount of calls into the call centre. Unfortunately the cost reductions intended were lost because they failed to manage the forum effectively. Customers now found their questions unanswered and the forum unwieldy and an aggravation.

Bad Sales

Here the supposed CX firm confuses hard sales growth with customer experience benefits. Sure in the short terms profits are made, but to the detriment of the customer experience and hence to the detriment of the firm.

- One large well-known management consultancy competed for a customer experience project with a Swiss insurer. Their focus was on hard gain to the company by focusing on the sales funnel. For them, answering call queries quicker would lead directly to more sales. The experience of the customer engaging with the insurer was, however, not considered. Picking up the phone quicker and pushing for sales is not the same as customer experience!
- A well-known telecom provider focused their efforts on customer experience by pushing sales campaigns. Great for marketing, but a campaign is not the same as the experience the customer has. And anyway, if mishandled through poor analytics, you risk sending out junk mail; in that situation a few will benefit but you can't tell from the sales figures alone the numbers that were annoyed, would have bought anyway or just switched off.
- An insurance company put on their website a 'ticker' to let customers know they only had 20 minutes to complete their transaction. The evidence was clear; there was a substantially better completion rate and revenues rose. Must be a customer experience then! However, no thought was put to the experience of doing this, nor the impact on other purchases: 'Oh I can't go and buy that insurance or spend more

time on the website because I have to complete this transaction.' Nor quite frankly the impact on the CX mindset of the firm. So the thought of getting people to complete was a good one; the experience of just providing a countdown ticker was not.

Notice in all this I am most certainly not saying upsell and cross-sell is bad, or process efficiency cannot create a 'noticeably easy' experience, one that does add CX value. All I am saying is, without considering the human effect at the end of the chain, you risk not doing customer experience.

Bad CX is Inside–Out

Overall, what these examples show is an inside–out view. An approach to CX that contradicts the core principle we outlined earlier; that is, focus on customer benefits not your own; use the source of value – experience – and increasingly interaction. Otherwise, you can only ever get a suboptimal CX result!

And of course, delivering efficiency only and a thin UX design you call CX, and encouraging bad behaviour opposed to the experience the customer has will only ever be a short-term approach because CX will not be in the DNA of the company and employees will become demoralised, that is, if they care. For instance, here is an example of 'inside–outness': the comparison being between the bad CX view of an IT supplier and the good CX view of its operator clients (Table 12.1).

In total, there was only a 23% agreement on the touchpoints of a customer experience between the technologist's view and the outside–in operators (source: three operators). Of course the main agreement was around the quantitative elements of finance and engineering, whereas the store, the main focus for decision making was hardly even considered by technologists as important to the customer. And no one considered interaction!

It's clearly a hard road to move your focus from yourself and what sells a box to the customer and what drives or could drive their motivation.

Table 12.1 Myopic inside–out view

Journey step	% Agreement of journey attributes, between inside–out and outside–in
Awareness	11
Marketing	40
Store	Less than 1
Activation	18
Recharging	0
Product	17
Website	25
Overall feel	0
Call centre	Less than 1
Network (use)	66
Billing and payment	25
Tariffs and pricing	77

Service Delivery Isn't CX

Which brings me to another point: service delivery is not the same as customer experience! If it was the same, we would not have a new paradigm and our management would be called service management. Let me explain by example.

In CX our starting point is to think of the customer as the main source of value to the business; our aim is to understand this source of value through greater appreciation of how customers think, feel and behave towards us (in other words their customer experience). Using this understanding, we then improve 'their' experience in such a way that they spend more with us over a potentially longer period of time and feel better about it!

Now a service delivery firm could rightly say service excellence is part of CX, when customers become aware of it, but on the whole the aim of service is to close the gap on expectations and optimise. Not to understand the totality of experience and engage in creating experiences that motivate and make money!

As an example consider how reliability of mobile phone connection is managed by operators.

This is quite clearly critical to maintain. However, its value in the mind of the customer is not as 'a subjective experience' but as an objective hygiene factor until of course it goes wrong. Hence, myopically assuming customers think only about connection speed 'as their experience' will lead the bad CX firm away from an appreciation of the totality of what customers value, that is, about the benefits that accrue from that connection (to talk to my friends) and even the totality of loss aversion, influenced as they are by cross-silo issues arising from poor billing and customer service.

Hence, mobile connection speed is not a customer experience paradigm but a service delivery one: most of the time! Its management is based on TQM principles and does not require mass customisation, collaboration and creative equity.

But if we take a CX point of view, the network becomes an experience platform and the question then becomes how can we engage the customers by for instance providing more OTT services or even by just understanding their perception of performance and working with billing! Which means of course, in its management CX requires a more collaborative innovative approach.

The Value of Effortlessness

Finally, I cannot get away from this topic without mentioning one other thing: effort. Now I like the book *The Effortless Experience*. Noticeably effortless is a CX principle. But for me, in spite of the talk, most of the time companies use effortlessness just to remove dissatisfaction. This is of course a noble service quality effort, but I feel the most value from effortlessness comes from when it becomes noticeable, an experience! There is no contradiction here; the effortless experience can provide CX value as has been proven by Apple. We don't just have to restrict effortlessness to hygiene.

Management Implications

1. CX is so bad because it is being defined through the lens of firm efficiency only.
2. There is a vast array of vendors, companies and organisations over-branding themselves with the term 'customer experience' without leading to any change in their ways of doing business.

3. For CTOs experience is a piece of machine data.
4. Where in all this are you throwing the Pike Place fish?
5. Experience is not about everything because not everything matters.
6. In the customer experience bad firm approaches include: a process efficiency focus only (which can contradict CX when the customer is not included) and bad sales growth.
7. There is a large difference in perception on what touchpoints are important between inside–out technologists and their outside–in clients.
8. CX has to be a relevant business strategy not a piecemeal activity.
9. Service delivery is only a part of 'the experience the customer has'. Bad CX firms just focus on expectations of delivery and do not consider the totality of experience nor its potentiality.
10. Customer effort is a good thing to focus on, but only in context. Effortlessness is a great way to create the Experience brand, as long as it is noticeably effortless, and adds CX value!

Part VII

And Finally

Customer experience is 'the experience the customer has'. This means it encompasses cognition, emotion and behaviour. So you cannot get away with talking about something being an 'experience' without the customer being in the loop.

So with that in mind let's look at how this is potentially resolving itself in today's market.

13

Interconnectedness

Our customer experience equation defined CX as follows:

$$E \times Q = f(PQ, SPQ, CJQ)$$

Hence, experience is a totality, the takeaway perception of customers based on a function of price attributes, service and product attributes and customer journey attributes.

So, focusing on price attributes as a competitive differentiator surely has an effect on what experience quality means to the customer. But it is a partial view. As it is if we focused on service and product attributes or customer journey attributes.

By contrast the advantage of the experience quality view is that when price is commoditised we can over-weight towards product or service; when this is commoditised we can look towards the customer journey. But the big insight is, if we change the E×Q to being something different and valuable we can revolutionise price and other aspects (as in the Starbucks example).

So the integral of this equation is potentially this balance between commoditisation and innovation. Defined as it is by the extent to which it creates customer value creation by creating a desirable goal.

But in what context does this apply? Do prior value demands (based on expectations) lead to what we notice as experience (as in I buy a can of Coke and that's all I experience) or does experience itself lead to or alter our definition of value? As we saw in the beach example: we wanted a nice beach, but experienced a qualitatively different experience which altered our value dimensions. Which incidentally is how experience has been used in SouthWest Airlines: make it fun to take their minds off the poor seating! Or to maintain premium pricing: I'll go to that dealership because they look after me better.

Overall though I don't think anyone has an answer other than 'Focus on how we connect CX to value creation,' not just experience for experience's sake. Or even emotion for emotion's sake; as we saw emotion is driven by 'the interruption of goal states,' how we appraise things in the light of our own well-being.

So always have the customer in mind!

But also be warned, while WE may identify something as 'an experience', if in the mind of the customer it goes unnoticed – except when it goes wrong – then most of the time it is NOT a customer experience: its value is zero. Of course, that doesn't mean we shouldn't control any potential risk of it becoming a damaging 'experience the customer has' but it is clear: with these types of experience we are dealing with a transaction not something that on its own will build a relationship.

For instance, today I am listening to a group of IT engineers talk about how they are managing machine thresholds and hence, the customer experience. I mean that's fine, but if you are only concerned with generating customer value through controlling loss aversion while the rest of the time customers don't even register this 'experience' I would say there is a contradiction in what you say.

Most of the time, if not all the time, what you are really managing is an objective dataset of experiences. Important, I agree, but still hygiene in the mind of the customer and certainly not relationship building.

In other words it's only an experience to the engineers looking at the information, not to the customer. And that way just leads to what Professor Stan Maklan gloriously calls 'the Theory of Everything.'

Likewise, we may identify something as 'an experience' such as a complaint, which is most definitely subjectively noticed by the customer. But this is already covered by principles of service delivery, so just focusing here is really only doing the same old same old and we must not conflate CX with service.

Therefore, we must differentiate CX as a separate source of value by calling out service delivery as say SERVQUAL, while still accepting they are fundamental platforms from which CX is built.

But I will be honest, most of the time companies remain focused on 'experience as everything,' the use of the term to mean the objective reality of things: customer not included. Which is a little sad. I mean wasn't the intent of customer experience to be about something subjective; the *experience the customer has*? Indeed I confirmed this in a quick straw poll of the various luminaries of 'customer experience' where there was a 100% consensus: experience is subjective.

Also, wasn't the intent of customer experience to be about responding to the commodisation of price, product and service by creating 'an experience', personal, memorable and maybe even a distinct economic offering?

Well yes!

Here is what Joe Pine has to say:

> Certainly all experiences are subjective; as I say, experiences happen inside of us in the reaction to the events that are staged in front of us. And certainly the dictionary definition of the term "experience" encompasses this idea. But when we single experiences out as a **distinct economic offering** …it has to be more than mere subjective experience of a good or service; it has to be distinct, which gets into engagement, memorability, and time well spent. In fact, I often use the formulation now, drawing a line between services and experiences on the Progression of Economic Value, saying that experiences are about time well spent, but services are about time well saved – there's the notion of convenience.

I for one find it ironic; customer experience was supposed to be a way out of commoditisation. Instead it's become commoditisation on steroids.

So how has this situation come about? It's almost as if firms are saying customer experience sounds sexy, there are some great examples of Experience brands, and we want to be an Experience brand, so what box should we buy?

For instance, I remember many years ago talking at the IQPC Telecommunications Conference in London. Towards the end it became apparent that every speaker was obsessed with KPIs: measure this, box that, experience, an undefinable and magical word for measure everything!

I couldn't quite put my finger on it but it seemed that the reason telcos are so bad at customer experience is precisely because of this obsession with metrics. They put efficiency-only KPIs before the consumer, and however important these things are, and they are important for internal alignment, we should not through this obsession, constrain, bottle and box a qualitative thing.

Hence, for me at least, in CEM the real mode of delivery has to be to manage a complex adaptive system: *the experience the customer has*. And that means the ability to flex and invent, understand where things are going, to think design and test the result! To seek upsell revenue, margin improvement relationship and engagement, not enforce a root-cause efficiency-only discipline of numbers that takes us farther away from understanding its qualitative nature. We need more balance and less abstraction from the C-suite.

A Short History of CX

But of course there is no doubt that customer experience started off from a position of good intent!

From the 1990s, fed up with zero defects and an undifferentiated logic of service delivery companies started to listen to leading thinkers such as Pine and Gilmour and later Vargo and Lush. Here, the 'customer experience itself" was perceived to be the source of value, not the product or service delivered. Hence, the customers' 'experience' of use and their personal subjective memory, their takeaway, was seen to extend opportunity beyond competing on the commoditised delivery of features and functions.

But theory is one thing, practice quite another. Hence from the 2000s we notice two diverse trends:

On the one hand we have organisations building on this subjective principle. Trying to answer the question, how can we create an 'experience' that is personal and memorable? Something customers would buy beyond product and service delivery (SERVQUAL), from which emotions derive.

Hence, we find Cranfield School of Management formalising experience in terms of deeper customer motivations. This led to the development of the Experience Quality dimensions, something I am proud to have written a paper on with Professor Hugh Wilson. Likewise, I tried myself to formalise emotion into a more standardised framework. Within this mould we also see the VoC vendors using enterprise feedback management and NPS, CSAT, CES and so forth.

But, these approaches have not led to better 'experiences' for the following reasons:

Reason 1: An inability to prove ROI

I see this problem manifesting in two ways:

Firstly, improving the *'experience the customer has'* often holds an indirect effect on customer value creation. Hence, unlike Opex reduction, its value is more difficult to quantify. I mean at an excellence level, if the call centre treats me well how can we know that such an intangible connects to why I do business with you in the future in a root-cause predictable way! Even though, as we have seen with the O2 example, hard direct causes never sit alone. This is the problem of modulators.

Secondly, CEM is a value-add. Hence, it depends as we have seen on innovation, which means taking the risk, a principle without guaranteed ROI. You only know after you have 'done'. This also means being aware of where the opportunities lie or might lie, which requires a degree of empathy, in other words understanding what it is to be a customer or employee, not what it is to look at a spreadsheet.

Reason 2: An inability to move away from TQM

A TQM approach is inconsistent with the subjective world of the customer which cannot be controlled! Your personal world is your own. The application of quality engineering thinking will only ever lead to a focus on cost cutting and firm efficiency, not the creation of experience.

Hence, we need a new focus on understanding how subjective experience works and on setting parameter targets, a very important principle from Cognitive Edge. (Parameter targets set boundaries: we need 'more stories like this, fewer stories like that.' This derives a fitness landscape that indicates where things are heading, its disposition from which we can take more realtime action.)

Reason 3: A lack of durable engagement

I believe companies have struggled to define which experiences are durable and valuable from those that are nondurable and valueless. Perhaps this is due to an inability to take the risk, or poor implementations that have not considered what drives the customer and its relationship to the brand. But nonetheless, this remains a problem.

I mean, Experience design is still a type of delivery, isn't it? So what happens once delivered? Customers will not necessarily feel engaged with you for the long term! So I experience the Theatre of the Geek Squad approach or the rapping air steward on Southwest Airlines: I like it; it reaches an unmet even emotional need; I go back. But through time, I am less interested in these experiences per se. Of course, I hedge my bets here as well; if the experience opens a new market need such as Starbucks' Third Place or Hotel Chocolat then durability has been reached; although of course this can be copied.

Hence, it seems to me, to resolve the durability problem we need less ethereal talk on emotions, as if they are somehow separate from cognition, and more talk on durable engagement and what this might mean in terms of value; hence, my emphasis on drives over memories (although of course one leads to the other!).

Olaf Hermans raises the point about lack of durability in the following way:

> The paradox of the experience economy is that it is impossible to stay engaged all the time and nor should you! You must allow for low engagement. For instance, once we have chosen a school for our children and got them in, we don't want to remain so heavily engaged. We acquiesce so long as they are happy.
>
> Emotions are not durable. We need to get to a cognitive state of durability, customers know what they get and can afford to engage in a lower mode of engagement.
>
> Experiences that are delivered do not lead to loyalty.
>
> Experience is good as a method not good as an engineered process. Any delivered experience does not make customers durably loyal.

I agree with Olaf, and have also said, that with loyalty and great CX we can see how emotional engagement is actually lowered into a more comfortable state through time. Or we say negative things because we care! Less emotion is not incompatible with loyalty or CX; hence we need to measure emotion and attitude better to make this apparent. And identify the drives and opportunities that lie beneath.

Reason 4: An inability to innovate

Firms have failed to innovate their ways of working. So although innovations are extracted from as-is measurements or exapted from people who live the experience or derived from their hunches, such as the Amazon hunch about packaging, firms are reluctant to take the risk.

Reason 5: A legacy environment

Failure has arisen because of the difficulties of achieving a customer orientation in a legacy environment focused on product silos. You see firms are built around functional silos not customers: how are you going to change that one! And why should you anyway? As I have mentioned before there are plenty of brands that make a fortune from this approach.

At the end of the day de-siloing may be too big a fish to fry, but is collaboration? Or sharing data through a single view of the customer?

Reason 6: A short-term mindset

The short-term KPI focus so dominant in the marketplace really works against the long-term benefits of customer experience and loyalty. Measurement, after all, if ill-used can create distance – it's your target not mine – which doesn't exactly help build engagement and innovation.

But taking my cue from the complexity movement, there may be a way to manage this if instead of planning for a mysterious abstracted long-term future or working for the short term we manage for dispositions. So we don't so much talk about 'a hypothetical world' or ignore the short-term requirements of business as work between the short and long term, seeing where things are going and to quote Dave Snowden, 'influencing the direction of travel.'

Moving Forward

Now normally these issues wouldn't be a problem. Some companies do it, some don't, but this use of the term experience as a subjective thing that encompasses aspects of service quality and goes beyond it is not the only definition.

Since the late 2000s faced with the inability to manage these points of failure a conflation has occurred.

The zeitgeist of IT, Six Sigma and Lean, and the critical rise of digital, has meant that experience has been redefined; now it's not so much about the customer but about the firm's concern for efficiency and cost cutting alone.

Experience as in the objective sense of the word! In this way, the customer's psychology can be ignored and only what we can objectively measure matters. What a fortunate thing indeed that the word is ambiguous!

Hence the experience pendulum has swung.

Customers are no longer assumed to respond to 'how they think or feel'; instead they are assumed to be interested in the nonexperience, zero defects, and firm efficiency: all things measurable by the firm, an industrial economy logic which of course means vendors scale their software and the nirvana of efficiency is the endgame.

Customer obsession means measurement efficiency obsession.

And now we are back to square one: a zero defects service quality logic!

So What Happens Next?

Fluid Thinking

I don't know, but one thing that strikes me from all these possibilities is that we need fluid thinking, the ability to create, combine and collaborate not focus on silos or efficiency, a necessity when all is digital and it's not the shifting of boxes that adds differential value or even the ability to scale but the capability to build relationships, innovate, mass customise and go with the flow.

Especially when the customer–firm interaction is not only king, but will run your business and when AI, quantum computing, big data and the Internet of Things will serve up immense opportunities for personalisation.

So for my money customer experience as practiced by many firms is now a suboptimal response. Its focus on industrial economy metrics – such as NPS, CSAT and CES – and failure to adapt to the fluidity of the fast digital economy has meant we need a new paradigm: that paradigm might be, for instance, Experience Design or a focus on customer interconnectedness (CI) wrapped around better measures of customer perception using complexity approaches, or the merger of objective analytics with subjective data to define 'situations' of importance to the customer.

Experience Design

Here designwise we would look at the broader 'experience' not just use. So, if perceptions of phone reliability are affected or potentially affected by bill shock we include that! Or if modulators are important we consider these.

Customer and Employee Interconnectedness

As we move to a more realtime relationship with the customer, intimacy and interaction between the firm and customer will become ever more important. Hence we will have to respond to and predict consumer drives as they arise, which will involve our understanding how consumers self-organise in the digital swamp and amongst their social milieu: one minute it's Facebook, the next it's Instagram.

Of course, we will still need to be efficient, and we will still find value from price, service and experience in the right context, but now we will be able to create new value through the interconnections we make between companies and between silos, such as the ability to personalise our reaction, giving the customer more control, making clear our intent to be with them in their journey, and humanising our technology.

This is of course driven by digitalisation and so-called big data, which will enable new ways to achieve customer intimacy as we saw in the ice on the roads example; we will also start to see how we and the customer can combine information in interesting ways!

Hence, the term interconnectedness! Where customer value comes from combining information flows not just interaction (which is 100% included but refers more to the end-communication).

So perhaps the real meaning of CEM rebooted is how we are moving towards new levels of *interconnectedness* and how the customer will define his or her own experience with our help!

Interconnectedness also means we will focus not on a single hit relationship as in 'I send you this discount' but on the development of a deeper understanding between sender and receiver, which is important because relevant and human interactions are the key to building a solid relationship, connections we will further use to promote our experiences and create higher levels of social engagement and sensitivity to innovation and market shifts.

Hence, in experience terms we can see that ultimately the customer does not have loyal feelings towards you – you're not a football team – they just want to know you have their best interests at heart and can create experiences that matter to them beyond just being efficient. From which loyal behaviours arise!

In this way, customer experience becomes about how you understand the customer continuously: your mindset, and how you collaborate and build relationships, without forgetting the critical role employee interconnectedness must take. The employee experience is also a source of value as well!

Excluding the term interconnectedness, this view is taken from the thinking of Olaf Hermans; I leave space for his words on the matter:

> The essence of my thinking is that companies can no longer compete on "delivery," albeit the delivery of product, service or experience. Instead, the basis of competition, and the only thing that will differentiate companies will be interaction.
>
> The unit of analysis then becomes the encounter, a "mini-project space" in which company and customer integrate resources. In this way, the customer delivers as much value (or more) as the company.
>
> In such an environment huge pressure is therefore put and questions asked of the "experience design" model. After all, what can you ever plan for before you meet the customer except information gathering and freeing capacity?
>
> Hence, loyalty is no longer "long term," it means "invested behaviour" and it happens here and now ("loyalty as a process" vs "loyalty as an outcome"): invest in the customer when the customer invests in your company, instantly.

For Olaf, then, loyalty is never a gift from the customer in the way it might be conceived in traditional CX. Instead, companies need to deserve it by interacting smarter, facilitate it by providing better information, and ensure provision of the right tools required by the customer to perform loyalty behaviour.

In addition, firms from a KPI perspective need to understand that NPS is NOT the goal! After NPS we need to facilitate loyalty in order to get it, this being critical today as brands are not able to support their staff in the encounter with such a large diverse customer base.

Olaf says:

Brands hope that the staff member is sufficiently empowered but are unable to grant time, budget, content and customer information to every staff member today in every encounter.

Employees are simply not able to behave in an empowered way, even if their supervisors tell them they should ("glass ceiling"). I predict that Google glass will be back soon as an essential part of triangle interaction: staff – customer – company/brand.

Loyalty then is no longer value based: companies, staff and customers, all have different sets of values, and will only be able to bridge their differences in value through interaction and meaningful encounters in which dialogue, not persuasion, is the communication form.

Value alignment is nonsense, value tolerance is king.

Olaf then raises some serious questions on the current CX approach:

Since Pine and Gilmore we have been focusing on customers' holistic impressions and emotions, a fine thing as customers don't want to be in 'quality mode,' but experience fanatics never solved the paradox that customers driven by emotions cannot be durably involved. **Involvement = cognition.**

Likewise, another problem is today, big data drives us back to Skinner: if we know what the customer does today, we know what he does tomorrow. Again this implies that the customer is numb and following patterns without thinking.

As far as I am concerned the last 15 years we thus have been living in the dark age of marketing in which we systematically denied customer cognition, the fact that customers are smart, make sense of what they feel, contemplate who they do business with and who else can do better, and especially think how they themselves can make the best contribution to value creation. This is not the same as thinking that they always act in a rational way. Cognition and being rational are not the same thing.

Luckily, the time of Artificial Intelligence has arrived which is the merger of (1) cognitive intelligence; (2) computational sciences. Companies now finally get assistance: (1) they can solve the problems caused by overemphasizing emotions and behavioral patterns and **promote dialogue** with the customer before making whatever assumption about the needs of the customer; (2) it will solve the problem of managers allocating resources to the wrong "personas" and customers and staff members, of generating all kinds of redundant content (campaigns, customer surveys etc), but also of forgetting to promote relationship formation and loyalty facilitation in every service touchpoint. Each contact moment should have 3 layers of base processes: functional, relational and loyalty processes.

So, here there are a number of critical concepts here worth repeating.

From Touchpoints to Encounters

Rather than consider touch points as the unit of analysis, consider this to be about the encounter. This is a mini-project where resources come together with the needs of the customer. For instance, an encounter is the 92 seconds at the hotel reception desk. Here, we have a black box; we have the knowledge of the staff member and the needs of the consumer, which can be variable.

Hence, an encounter cannot be designed for, only prepared for between both parties. And over the course of the encounter, expectations will change.

In the encounter there will be a set of needs related to the relationship. Which doesn't mean we want engagement all the time, many times we just want to be in lazy mode. After all low engagement does not mean low loyalty.

Employees over Brand

Experience has previously conceived quality as a triangle between customer – employee and company. Hence, the company has a brand promise; customers have no expectations and follow this promise. This in turn puts pressure on the delivery system through compliance and guidelines. The employees are seen to represent the brand.

In the encounter model, everyone represents themselves! Hence, the question becomes how the brand can facilitate the encounter; for instance through information and resource integration. This is not symbolic, or persona based, this comes down to the conversation. We need in essence to ground brands. The standardisation here means briefing and de-briefing.

Loyalty and AI

To create the necessary conditions to engender loyalty we need all operational systems to be grounded in feedback. This need not be highly abstract but more in the nature of 'do you feel important' 'are staff happy and motivated'. We need to put genuine interests at the core of business.

This is too complicated for managers to manage! No need for abstract personas. No need for scrum. Instead use AI to support invested behaviours. I mean who are you to say we should not invest in behaviours with a first time customer just because they have not demonstrated loyalty or only invest in them after the twelfth encounter.

At the end of the day customers are looking for simple products, services and brands. However, relationships are more complex because they are more durable. So loyalty rewards are only 15% of a relationship.

The essence of the relationship is I do stuff for you; you do stuff for me. AI can model this, hence we need tracking. Start with feedback, where the content is delivered at the right moment to the right person in the right way within the relationship context. Not 200 questions, but single questions of relevance.

The quality movement want a single question but convenience does not mean relevance. We need more questions in the encounter. Collected in a smart way; in this way in all environments we can make the encounter richer.

Three Types of Need

The functional need to achieve efficiency and effectiveness – I give you your room key.

The personal engagement – how are your kids doing

The working customer – can you do me a favour and check-out at this time, due to the long queue: since there is more pleasure in giving than taking.

Ironically in loyalty there is the concept of the working customer, someone who makes a contribution to the encounter. When customers make their own contribution they are more satisfied with you as an organisation. Hence, we need to make the connections stronger.

This goes beyond the value-in-use concept of Experience.

Get Grounded

We also shouldn't seek to over abstract things. Customers certainly don't. Instead we should take a more grounded and down-to-earth view. Take the example of loyalty. Your neighbour would not understand you saying you are an ambassador for loyalty. They would just look at your invested behaviours, how you 'do something for them'. Hence, loyalty is a process an activity.

Customer Value Creation

Going forward then we must connect the notion of experience to customer value creation, which means we cannot focus only on the transactional or hygienic. These things just set the platform for the relationship.

We must also go beyond SERVQUAL, beyond the belief that everything is about managing how customers calculate expectations against ratings of reliability, assurance, tangibles, empathy and responsiveness. This is a limited approach to the phenomenology of experience where customers:

1. Construct meaning as it comes to them consciously, emotionally and nonconsciously.
2. Do not use a ruler of expectations all the time.
3. Find new information that is meaningful to them and qualitatively different.
4. Are affected by benefits sought.
5. Respond to other elements of brand, value through time and cross-silo/ cross-journey aspects (e.g., how network perceptions are affected by price, and NatWest mortgages by behaviour of NatWest car insurance or Morgan Sindall onsite delivery by the standards set by Mandarin Oriental), the social environment, aesthetics and peripheral clues.

All of which leads not to customer reality management, but a need to define what is meaningful and hence what is or could be valuable.

So, at a transactional level when I use my mobile phone, what is 'of value' is whether it works. But this is a given (for now!), and represents business as usual. Hence, what is also 'of value' could be:

- The benefit that accrues from my call or data use; in other words, my *value-in-use*: 'I get to speak to my family'; 'I keep in touch with the football scores.'
- How the handset works.
- My impressions of the brand from others.

Value also includes *fleeting* affects; the nonconscious as well as the potential that lies in targeting the customer's deeper *drives*; such as the moment I suddenly got crystal-clear sound quality on my handset due to software on my handset; or 'care' moments when the company expressed concern at my high data usage.

These *fleeting* things are important because they change the background noise of my opinion.

In addition, there is *value-in-relationship*, which highlights other aspects of the experience beyond value-in-use.

For instance, how you send me spam emails to buy or top-up when I don't want to be bothered by you. In value-in-use terms these things are unfortunate, in value-in-relationship terms, you are trying to rip me off! I feel it personally. I don't just see myself seeking benefit from you but also how you personally treat me!

And this is particularly important in the sensitive buying decision.

Here, for instance, I may say it's the phone connection that's important but in reality I see it as a given between providers. Hence, value creation is about how you relate to me, the information given in a store or online and the way it was given. The interest you elicited from how you retail your service lines, how you have treated me over the long-term course of the contract, how you persevere with me and so forth are all things that indicate if I should trust you for the long term.

Experience value then is not just a fixed thing but also more complex. We construct experience each time more than we adjudge things against a ruler of expectations.

Which means that most of what is of value is new information received by the customer, information that is qualitatively different (as we saw in the example of the different type of beach), or slightly amended each time we visit (as we saw in the Charles Clinkard store visit). Hence, when we think of value we need measures that are less static and more dispositional, narrative-based, flowing and immersive; more in tune with the nature of subjectivity.

Understand Value

Hence, it is my belief that we can get better insight through sense-making and approaches that make the personal phenomenological world more transparent. Something statistics and calculative approaches mostly fail to do because they just reiterate the functional in their application and fail to understand that customers mostly 'don't' calculate.

C-suite executives must understand that customer value is more than the functional otherwise we will forever play a commodity game and miss the customer!

Hence, mindset change is critical:

If we can get a better understanding of value then we can stop positioning roles as only functional, for instance how HR focuses on measured functional activities rather than on creating employee value.

Get to the Question

However, I still believe that before we get to the answer we need to first consider the question.

Too much for me has already been written about grand schemes to de-silo the organisation, customer obsession, put the customer at the heart of things and so forth. But none of this can be done if we still seek to KPI in a manner that creates dysfunction, distance, an inability to change and a failure to understand the customer.

Perhaps you will now understand why in this book I have put so much emphasis on measurement and the use of complexity tools and other multimethod approaches and the need to set parameters rather than targets, as well as the need to get down to brass tacks and seek to re-understand what it is to say: 'We want to do CEM.'

Hence, I believe that raising the efficiency–experience debate is crucial.

Which of course asks the question of leaders: which way do you want to turn? What outcomes, and behaviours do you seek?

After all, what's the ROI on customer experience? Survival! You have to move with the market and evolve.

Customer experience is a business strategy, not a box. If others are doing it and you are not then you will end up myopically focusing on Opex reduction, leaving yourself even less able to invest to keep pace, let alone grow. For me, you have to deliver CX consistently and through a shared vision or purpose to quote Shaun Smith, while remaining flexible enough to respond to the changing environment.

After all, uber-efficiency only leads to over-specialisation, a myopic inability to scan the environment and the threat of finding ourselves old fashioned and out of date as markets and competitors change exponentially with the digital revolution!

So, don't fall into the trap of thinking objective efficiency for the firm is the same as subjective experience or that CX is a term without utility just because the word 'experience' is ambiguous and can be used as a marketing ploy to scale anything. For me, that's customer reality management not customer experience management.

I end then with a quote from Michael Young, CTO and CEM expert:

> CX provides an opportunity to break the normal sales cycles and provide guaranteed longevity on a scale that can't currently be achieved with the bits and pieces approach…… it's the "combo meal" rather than the "big mac" ……..

So, are you an Experience Brand or an Efficiency Brand?

Don't wait too long to make your decision.

The digital revolution necessitates we rethink CEM and how agile we are as a company. Not least because it means businesses need to embrace and enable innovation, agility and an ability to evolve. Not focus on firm efficiency alone: that is a recipe for glacial slowness, something you might have been able get away with in the old bricks and mortar world where fast brand turnover was less resonant.

Now we need to get close to the customer; we need to create customer experiences that motivate and engage. We need to create customer value.

Here Then I End with Some Comments from Michael Young

Work Cross-Silo

For the last 20–30 years I have been around technology. In my role within Hutchinson Telecom, my CEO gave me the role of combining both CTO (Chief Technology Officer and Customer Service). He could see that this would lead to a better result for the customer since the best result for resolving network and product issues is to be had by breaking down the silos and having a single organisation. The aim here, then, was to make sure we responded to issues early and proactively, combining teams with collective accountability.

This meant changing the group culturally.

To make the organisation ready, we had to understand that customer service and technology think differently. So, people need to be open to learning and understanding each other's challenges while being focused on the outcome of service.

Hence, we had customer service and technology all sitting around the same table and investing in areas they need to rather than seeing issues siloed. To do this you have to go with an open mind, learn from each other, challenge your thinking, understand and respond. Do not go in with preconceived notions based on just your world view.

It's the same with managing network NPS, its not just the responsibility of the network operations team; the promise of performance comes from marketing; so you can have a fantastic network but it is sold in a different way. Hence, you must work cross-silo.

Michael Then Describes the Importance of Customer Focus...

When I left school I was a cocktail barman. What I learnt was that it wasn't about the product or the service, it was about the combination. In CX it is similar. The customer brings something to the party you have no control over. So if they are in a great mood and you mix a good cocktail with good conversation you get a good time. If they are in a bad mood, and you approach them in the same way, you get a different experience.

During my time, I couldn't work out why I didn't get more chips than the others: it was the fact that you can make the best product but if you didn't put the effort into the experience then you could not guarantee a good result.

What This Means for Metrics

This also means that it is important to consider not just one measure such as NPS but a cocktail suited to customers and organisations.

So, one of my frustrations with CX is that you can have a metric such as NPS but it won't tell you about the experience. To illustrate, in one exercise I interviewed 70 people as well as used metrics such as NPS and CSAT. What I found was that the stories customers tell did not match the number. So numbers might be 6 or 7 out of 10 but the story we were getting back did not match. Now if I had put the number in front of the board it would be nowhere near as confronting as those stories. Stories show up some of the ridiculousness of numbers.

I am not advocating not using NPS, just that organisations that say there is a single metric make me want to pull my hair out. Numbers can hide deep-seated problems hidden in the small stuff; not the 98% reliability figure.

And we can extend this to what metrics work best based on the type of company. So with one organisation I was dealing with where employees where very young, what worked was emojis not a 0–10 scale: you don't come out of a movie saying that was 5 out of 10.

What This Means for Design

Smart designers should be more involved in decision-making; start to engage the customer more flexibly in order to get customers bound into the decision. Fit in with their organisation as well as their customers.

I get frustrated with the box sellers. Everyone is in a different place; it's not a one size fits all. You have got to see what the needs are and what the company is capable of. There is no point being an Arsenal if you are a Southampton.

And you need CX hard-wired in your company to evolve.

5G is an example. In 5 years there will be no data plan; it will be commoditised like voice. The data package will disappear. You need to sell ahead of the curve. Otherwise you will get destroyed, since customers are becoming more fickle, accelerated by the ability to move.

CX then is a huge missing opportunity that builds out the fabric of 5G; the opportunity to create a difference. So why isn't it being done: would Apple have a problem mandating an extra % spent on innovation? It's not the computing power, but what you do with it, after all we sent a man to the moon based on the power of a calculator.

The Criticality of Brand

Companies need to do due diligence on what they are trying to achieve. Understand what the company is capable of and what the customer is looking for.

You need to create companies that can respond much quicker. Especially as information accelerates and gets out to the market.

Sell with RnD

From a technology point of view, get a team of CX people to work with the engineers in RnD, spend time to add a layer of time and attention to things people are not thinking about.

Companies have to be open, listen and understand not preach then come back with an interactive way rather than look to sell another product.

ROI

You are not going to find ROI – but if you tell me what you are looking for from customers and the brand then the money will take care of itself. There is no difference in measuring the ROI of training staff than CX. But this is not a leap of faith, you can draw a line between CX improvements and finances without breaking this down to $; if you go down one path and it is better for customers, and you are consistent with the brand then your will be better than the competition.

NPS has overstated the relationship to the health of finances for so long; but it's not the case all the time.

> To me CX is still the most interesting thing happening/not happening today, fascinating!

Footnotes

[1] A cognitive assessment of experience is about intangible benefits as well as tangible. That is why a focus just on the tangible such as 'website', 'network performance' or 'booking' misses the point that customers see the world qualitatively, influenced by their subjective senses, based on 'feel' together with reason. Examples of this include their sense of trust, how satisfied they are, the 'feel' of the brand and so forth.

[2] An objective-only definition will focus investments on technology, which is a considerable cost in terms of asset base if not connected to customer value. In this way, it may constrain opportunity-seeking behaviour, the ability to look for frequently small effective changes in the experience that mean something to the consumer. So a smooth tech platform that works, while being very important, would not on its own deliver a LUSH or Metro Bank experience. Nor would it lead to better people engagement.

A technology-focused environment also risks creating an obsessive focus on risk management and hygiene factors with 'operators' confusing machine data with 'that must be how the customer thinks.'

Hence, we lose the capability to focus on intangible benefits, the very thing consumers buy. Tangibility is only important as an enabler of intangible benefits.

[3] This is what happens to many marketing concepts. You can't scale an idea or expertise, but you can scale software. So, if we take the experience the customer has, then feed it through the mincer, TQMing it, breaking it down into components and in the process miss out on how cognition really works, we can measure and monitor it. And voila we have a practice. This has the unfortunate effect of giving us false hope while at the same time failing to account for experience losses.

[4] Customer experience has become a general term for: (1) service design, (2) seeing things through the customer's eyes and (3) serving up any old thing as 'selling experience'. Apart from the last item, this is not a bad thing. But it does mean it holds little difference in meaning from say service and customer research! Hence, service design should really be experience design.

In addition, the complexity of digitalisation and multichannel integration has meant that the words 'customer experience' are just useful terms with which to talk about user experience. This is all fine, but an over-brand is not enough, even if it does give a sense of mystique (not quite definable, slippery, can mean what you want, magical!). We need to consider the experience economy definition and how we can create an experience not manage one.

[5] I am aware that relationship marketing is the origin here and that there is a clear ROI of CRM based on software. However, I stand by my comment: if you use a qualitative term like relationship then you need to demonstrate a qualitative benefit!

[6] If you take a technology enabler to be 'the customer experience' you risk assuming that when the machine KPIs work well then the customer must be happy! This is not necessarily the case if the customer thinks of the 'enablement' as hygienic. And certainly not the case because no customer buys the tech enablement; they buy the benefit that's derived.

Likewise, where do you stop your technology enablement as an experience definition?

A back-office document bill printer has a critical role to play in the experience of electricity and gas buying. Is this now a customer experience solution?

[7] Hence the old adage that claimed drives only weakly correlates to spend behaviour is true, but misses the point!

[8] Taken from Dave Snowden of Cognitive Edge.

[9] I avoid the transformational experience which goes beyond customer experience. But point the reader to the book *The Experience Economy*, by Pine and Gilmour, to understand what this means.

[10] Of course we can see drives strike across all economic areas, but it's the deliberateness of the approach that is important. Assurance seeks to assure, Excellence seeks to excel, Drives seeks to look deeper at what is beyond the current offer.

[11] Fix is often conflated with process efficiency and cost-cutting, but you can merge the two, so customer experience is a safety valve for Opex reduction, ensuring that any cost-cutting does not chop out value and acts as a safety valve for sales growth, so that sales growth leaves the experience at least unchanged. But in an ideal world CE is your lever of growth.

In addition, a focus on hygienic process efficiency can release time to engage the customer so even here they are not mutually exclusive. The retail firm that 'Apple like' gets their shop assistants to take payments via tablets means they also have more time to talk to customers and engage them in the products on sale. As long as other customers are waited on and don't see a greeter standing around!

[12] I call these Type 1 and Type 2 rather than put them into distinct categories because they are both about deeper engagement and drive. Academics call this value-in-use

[14] The paper, 'Network as an Experience Platform' (Source: Pine, Walden, McCann-Murphy, 2013) emphasises this point.

[15] Sure you get different types of bolt, but it's hoped you appreciate the qualitative difference.

[16] For instance, if you look at customer stories around a score such as 8 out of 10 on NPS you will find a vast difference in emotion words used and drives expressed about the score and around how customers perceive you.

[17] This example shows the general effect. I accept some goods such as luxury products may flatline at a higher score, but the point is the data are curvilinear and not about 'more of' a score all the way to 10 delivering higher growth. As we show even 10 out of 10 is not consistent in meaning! Note: The curve is referenced in part to Professor Stan Maklan and Cranfield School of Management, with my take on it.

[18] I also note that there are other issues such as the validity of customers responding unnaturally to a survey and how representative this is of the population who aren't survey primed.

[19] Stories must be picked up relevant to the situation. In other words not just an abstract 'Tell us your story' question, but 'Tell us your story related to purchasing a phone, recommending a friend and so forth.'

[20] Being oblique in questioning is also important. Too constrained and you guide response, too open and responses become too abstract.

[21] The irony here is that NPS does not measure dissatisfaction and a bipolar CSAT scale means consumers answer towards the neutral because a call centre has aspects of satisfaction and dissatisfaction, or answers tend towards the positive because you ask me about satisfaction! And I gift you an answer.

[22] 'There is evidence that people spontaneously regulate their emotions (Forgas and Ciarrochi, 2002). Immediately after an emotional event, people in both happy and sad moods experience more mood-congruent than mood-incongruent thoughts. With time, however, the content of people's thoughts moves towards the opposite valence. That is, after a few minutes, participants induced to feel sad were having happy thoughts, whereas those put into a happy mood had relatively more sad thoughts. This homeostatic emotion regulation fits nicely with the current analysis: mood-congruent thoughts help people learn the lessons of their previous behaviour, but adaptive future behaviour requires that emotion regulation take place.

[23] You should seek more of something if you are a poor performer. If you score a poor 5 out of 10 then you should try to raise your score. However, if you are close to a plateau, say you score 8 out of 10, then expect diminishing returns from a 'more of' approach! However, there is another option. Rather than try to achieve a higher figure, seek to change the meaning of that figure. We could make a score more resilient. So, network reliability may score 8 out of 10 on NPS today because the brand is well known and it might score 8 out of 10 tomorrow but be more associated with network quality! And hence drive more value. Likewise, rather than get shoppers to go to a grocery store and score 9 out

of 10, accept that they score 8 out of 10 but make the meaning of 8 out of 10 deliver more value. For instance, 'I' go to the store and the product I wanted is on the shelf, but now it has accessories. It's not that I score more; it's just the same score is now more resilient because you offer more of the product I want. Or 'I' go to the store and I see a cooking display. I still score the store 8 out of 10 but there is more resilience in my association of the store with cooking products.

[24]Indeed, one of the claimed best examples of customer experience comes from the Royal Bank of Scotland. Here they brought in an engineer to manage the programme who focused wholly on cost reduction, Lean and Six Sigma. The objective became not 'What can we do for the customer to increase the subjective asset, customer experience,' but 'How can we benefit ourselves, cutting out processes that seem superfluous.'

The case study was a great success but was it from a customer experience point of view? Yes, it reduced cost but my question is: Did it deliver an appreciable return on subjective experience? Furthermore, who is to say that some of these invisible cuts might well have been important components of a customer's 'experience,' or set the platform for more experience creation? I would contest that without a valid multimethod measure of experience no one could tell. But its use of activity-based costing is a significant addition to the CEM literature.

Other Comments

Embedded value refers to the product and service attributes, the value-in-use refers to the outcome, purpose or objective served in the product or services consumption.

Holistic: Not a great word because it just brings us back to 'sum of everything,' definitions which I reject.

Resonance: This is similar to the concept of building memory assets that Professor Byron Sharp talks about in his book, *How Brands Grow*. Here we are building memories that are deeper and more personal which leads to a higher level of commitment. Resonance for me also has nonconscious as well as conscious aspects, the phenomenology of what resonates with me.

Emergence: There is no ROI on 'look and feel' at least not one that can be easily predicted beforehand.

Resilience improvements: The improvement in resilience is measured by the change in customer story associations not in a change in say an NPS number. There are, for instance, more positive stories and fewer negative associated with the number. One thing that might help is a survey process also for employee stories, continuous feedback on the environment to look for de-resilience. This is not the same as surveys that serve up set items.

Drives: These do cut across all areas, but are a better word than emotions because they lead us to think deeper about the customers, that is, how they think, feel and behave, both consciously and nonconsciously.

Commoditisation: This is a relative not an absolute condition. Hence there is constant cycling, efficiency–excellence–drives back to assurance. In parallel!

Dissatisfaction: Dissatisfaction is an important and underrated measure. After all, it actively gets to negative moments! You can see how valuable this might be in situations where it actively teases out dissatisfiers that might be hidden in a bipolar satisfaction scale.

Hence, if you asked me to rate my level of satisfaction with a call centre I might give it a middling neutral score (between satisfied and dissatisfied on the same scale). Why? Because there are some parts I am satisfied with and some I am dissatisfied with; essentially by not actively asking for dissatisfiers, these have been muted.

Special Thanks

Dr Simon Moore
Dave Snowden
Joe Pine
Olaf Hermans
Tony Quinlan
Professor Hugh Wilson
Dr Emma McDonald
Dr Nigel Marlow
Dr Katharina Wittgens
Phil Lewis
Michael Young
Gautam Mahajan
Rob Frank
Harry Cole
Paul Ryan
David Pinder
Christopher Humphries
Shaun Smith

Anyone else I have forgotten to mention but contributed to my appreciation of customer experience!

© The Author(s) 2017
S. Walden, *Customer Experience Management Rebooted*,
DOI 10.1057/978-1-349-94905-2

Index[1]

A

AI. *See* Artificial Intelligence (AI)
Amazon, 1, 22, 65–6, 86, 133, 169, 225
Amex brand, 131, 152
analytical equity, 135, 137, 199, 201
anthropologists, 200
Apple, 20, 31, 35, 54, 56, 86–7, 91, 126, 157, 160, 169, 172, 174, 215, 239, 243
arousal (high and low stress), 155
Artificial Intelligence (AI), 15, 22, 230, 232
assumed certainty, in emotions, 155
attitudes, 24, 133, 157–8, 165, 225
attribution, 174

B

BA flight, 124
balance sheet, 199
behavior, 133, 136
behavioral patterns, 230
behavioural psychologists, 175, 200
Bezos, Jeff, 133
BIC, 54
Big E approach
    customer experience approaches, 59, 60
    financial success and experience, 58–9
    personal and memorable experiences, 57–8
    transformation power, 58

---

[1] Note: Page numbers with "n" denote notes.

© The Author(s) 2017
S. Walden, *Customer Experience Management Rebooted*,
DOI 10.1057/978-1-349-94905-2

Index

bill printer, 242n6
Black Horse restaurant, in Woburn, 132
blue dot effect
   customers gift scores, 105
   NPS scores, 104
   objective statistical model, 105
   regression techniques, 104
brand identity, 143
bread-buying experience, 79–80
British Airways, 131
Broadstairs, 115, 132
Build-A-Bear, 199
business intelligence, 139

C

call centre example, 110
campaign management, 22, 208
CAPEX investments, 147, 150
Capital One, 199
Caribou Coffee, 107, 141
car repairs, 117
cause-effect relationships, 109, 114
CEM. *See* customer experience management (CEM)
Cerritos Library, 41, 58
CES, 100, 223, 227
Chat Live!, 43
Chief Technology Officer and Customer Service (CTO), 28–30, 34, 85, 141, 187–8, 208–210, 236–7
churn analytics programme, 144, 208
churn rates, 144
CI. *See* customer interconnectedness
CIO, stakeholder, 138
CJQ value-in-use, 65
client interests, database of, 197

Clinkard, Charles, 116–17, 136–7, 173
cognition-involvement, 230
cognitive assessment, 1, 127, 241n1
Cognitive Edge, 89, 109, 130, 242n8
   principles, 132–3, 224
cognitive intelligence, 230
Cole, Harry, 247
commoditisation, 220, 221, 246
commodity game, 235
competitive differentiator, 25, 35, 65, 219
complaints, 15, 28, 42–3, 47, 86, 97, 139, 162–3, 188–9
complex thinking
   cause-effect relationships, 114
   Charles Clinkard experience, 116–17
   cost-cutting activities, 118–119
   employee system, 116
   flow of experience, 114–15
   habitual experience, 117–18
   resilience management, 115–16
   resonance, 117
   trial and test, 117–19
computational sciences, 230
contact centre software, 208
contract relationship–Perfect Delivery, 198
cost-benefit calculation, 23, 45, 77, 107, 154–5, 188
cost-cutting activities, 118–19, 224, 226, 243n11
cost reduction, 210, 212, 245n24
costs savings, 199–100
Cranfield School of Management, 53–4, 139, 223, 243n17
creative equity, 135, 137

Index

CRM. *See* customer relationship management (CRM)
cross-silo innovation platforms, 137
CSAT scale, 223, 227, 238, 244n21
CTO. *See* Chief Technology Officer and Customer Service
customer
   executive meeting trap, 30
   and firm account, 34
   myopia trap, 30
   tangible trap, 31
customer advocates, 192–3
customer-based differentiation, 35–6
customer care, sense of, 35
customer differentiator
   equation, 64–5
   examples of, 50–1
   personal gains, excellence, 51
   service and product efficiency, 42–3, 49, 53
   SERVQUAL (service quality), 49
customer drives
   business executive or management consultant, 17
   customer psychology and expertise, 17
   direct and indirect relationship to behaviour, 13–14
   emotional dispositions, 17
   emotional value, 14–16
   enterprise feedback management, 14
   intangible drives, impact of, 14
customer effort score, 208
customer experience
   bad sales, 212–13
   complex thinking, 114–19
   cost-benefit calculators, 107–8
customer relationship management, 4
CX or CEM, 5–6, 10
definition, 10–13, 207
drives, impact of (*see* (customer drives))
emotional value, 3
experience management, 12, 21–2
firm efficiency, 211–12
inside–out view, 213–14
market research, 11
NPS question, 109–13
objective analytics, 109
personal and memorable experiences, 3, 11, 13
qualitative difference, 4–5
sensory information, consolidating, 9
statistical root-cause approach, 108
customer experience equation, 35, 64–5, 219
customer experience management (CEM), 208, 210, 222, 223, 236
AI (artificial intelligence), 22
attitudinal metrics, 24–5
CEM tools, 22–3
competitive differentiator, 25
customer service, 25
Efficiency brands, 21–2
Experience brand, 28
'experience improvements', 29, 32
Goodhart's rule, 23
governance process, 28–9
handling customer subjectivity, 26
innovative space and cross-functional teams, 27

customer experience management (CEM) (*Cont.*)
  'inside–out', 22
  personal, memorable and targeted experiences, 26–7
  physical things, 24
  qualitative and personal information, 25
  summing all experiences, 23–4
customer experience (CX) methods
  Cognitive Edge principles, 132
  effective journey mapping, 134–43
  hotspot narratives, 132
  human reasoning, 133
  inside-out layer, 143–6
  management implications, 149–50
  narrative metrics, use of, 130–4
  shareability, advantage of, 131
  weak measurements, 130
  web browsing speed, 129
customer experience, return on
  commoditisation of excellence, 38
  drive, 38
  efficiency, 38
  excellence, 38
  product–service–experience, 39–40
  subjective asset management, 38–9
  types of experience, 37
customer–firm interaction, 227
customer focus, 238
customer interconnectedness (CI), 227
customer journey attributes, 219
customer journey quality (CJQ), 25, 54, 64
customer motivations, 223

customer narratives, 130, 163
customer obsession, 227, 236
customer relationship management (CRM), 4, 10, 207
customer service, 25, 43, 80, 112, 132, 156, 172, 194, 201, 208, 215, 237
'Customer Talk Test', 60
customer value, 235
customer value creation, 220, 223
  calculate expectations, 233
  fleeting affects, 234
  limited approach to, 233
  value-in-relationship, 234
  value-in-use, 234
Cynefin, 109, 141

D

Damasio, Antonio, 151, 154, 170
data
  objective, 75–6
  subjective (*see* (subjective data))
decision making moment, 133, 134, 154–6, 164, 179–81, 239
defeats, 131
degree of empathy, 223
Delta Airlines, 140
DEM model, 18
  drives, 19
  experience layer, 19
  feedback loop, 20
  memory layer, 19
  'Real' CX, 20–1
depth interviewing, 139
depths, 130
desires, 131
DHL, 199

digitalisation, 65, 242n4
Disney, 1, 58, 178, 194
disposition, emotional characteristic
   anticipated emotional reward,
      levels of, 160
   environmental mood, 162
   framework, 179–80
   peripheral clues, 161
   story metrics, 163
   value-in-use frameworks, 163–4
dissatisfaction, 44, 215, 246
Doubletree Hotel, 148
dreams, 131
drives, 246
Dubai Shopping Mall, 147
durable engagement, in customer experience, 224–5
Dutch telco, UK financial services provider/global transportation company, 195

E

Edge, Cognitive, 130
efficiency benefits (FX), 35
The Effortless Experience, 215
Embedded value, 245
emergence, 246
Emirates, 140, 148
Emirates Airlines, 50
emotion
   attribution, 174
   bio-connectedness, 181
   cognitive agitation, of nervous system, 156
   conditions of 'assumed' certainty, 155
   conditions of uncertainty, 154
   consumer psychology, 169
   decision-making moment, 154
   design for, 169
   emotional drives, identify of, 169
   fleeting by nature, 156
   informational feedback loop, 154
   interruption of goal states, 220
   loss aversion, avoidance of, 170–1, 183
   managing stress, 173
   memorable moments creation, 173
   mood effects, 171–2
   negative emotion, 157
   nonconscious effects, 173
   Peak-end rule, 175
   peripheral clues, 161
   personalisation, 171
   rational thinking and social behaviour, 154
   resilience, 182
   right experiences, 165
   social environment, 171
   state of alertness, 154
   weak signals, 160, 182
emotional desires, 155
emotional disposition, 17, 160
emotional fears, 155
emotional impressions, 130
emotional intensity score, 158
emotional landscaping, 138
emotional omission, 177
emotional reward, 155
emotional statements, 157
emotional tags, 170
emotional values, 61
emotion theory, 12
emotion words, 158

Emotix study, 141, 159–60
employee and customer,
    interconnectedness
  Artificial Intelligence, 230
  customer value creation, 233–5
  digitalisation, big data, 228
  employee experience, 229
  functional need, 232
  information flows, 228
  loyalty, 229
  realtime relationship, 228
  sender and receiver,
      understanding
      between, 228
  working cross-silo, 237–8
encounter model
  between both parties, 231
enterprise feedback management
    (EFML), 14, 223
evangelist network, 200
exaptation, 132
existing surveys, 139
'experience as everything', 221
Experience brands, 222
  concept, 28, 31
  customer differentiator, 42–4
  customer experience, 45–7
  efficiency, 48
  examples of, 41
  fixation on fix, 47–8
  ROI, 44–5
  SAM approach, 41–2
  values, 191–2
Experience Design model, 143–4,
    224, 227, 229
The Experience Economy, 9, 243n9
Experience Hubs, 193
Experience Quality dimensions,
    40, 223

experience towards
    transformations, 60–1
experience *vs.* efficiency
  cognitive assessment, 1, 4
  personal and memorable
      experiences, 2–3
expert judgment, 130
ExQ (experience quality), 25

F
Fabergé, 90, 159
Facebook, 228
fears, 131
feedback loop, 118
firm efficiency, 224–7
First Direct, 41
fleetingness, in emotional
    characteristics
  negative moment, 156
  nonconscious, 158, 182
  relationship surface feature,
      158, 182
  sentiment text algorithm, 157
fluid thinking, 227
Focus on Do, 144–6
Frank, Rob, 197, 247

G
Gatwick Airport, 148
Geek Squad Theatre, 57, 224
Gertie, Golfing, 27, 67
Giff Gaff, 137
Gilmour, James, 9, 11, 222, 243n9
glass ceiling, 230
goal-state mapping, 175
Goodhart's rule, 23
grey experience, 141

H

Harrods' food quality, 195
Heathrow airport, 167
Hermans, Olaf, 11, 63, 133, 225, 229, 247
hidden drives and uncover deeper drives, 139–40
histograms, 131
holistic design thinking, 178
Home Alone 2, 9
homeostatic regulation, 157
Homo narrans (storytelling person), 131
Homo sapiens, 131
Hotel Chocolat, 56, 224
hotel trade, 27
hot-house respondents, 138
hotspot narratives, 132
How Brands Grow, 245
human behaviour, 130
human brain, 129
human interactions, 228
human modes, 133
human reasoning, 133
Humphries, Christopher, 247
Hutchinson Telecom, 237

I

immersion, 130, 138–9
industrial economy metrics, 227
information quality (InfQ)
  data sharing and analytics, 66–7
  early stage, 68–70
  efficiency towards experience, 69, 70
  personal and memorable 'interaction' experiences, 66
  'personal appointment', 69
  process efficiency, cost saving and Opex reduction, 65, 68
  self-expression, 68
  social communities and microsegments, 67–8
InfQ. *See* information quality (InfQ)
innovation, 220, 223, 225, 228
Innovation Bubble, 48, 138, 141
Instagram, 228
Institute of Customer Experience, 208
Institute of Customer Service, 208
Internet of Things, 227
interpreting intelligence, 133, 134
invested behaviour, 229
IQPC Telecommunications Conference, in London, 222
IVR system, 126

J

Jaguar Land Rover, 193
Jazz Café, 86–7
JCB, 199
journey mapping method
  analytical equity, 135
  beautiful design, 147–8
  blank sheet of paper, 135
  ideation methods, 137–41
  journey maps, 134, 149
  listening, best way, 135
  open listening, 135
  purpose, constrain things, 136
  touchpoint quality expectations, 146
  value judgments, 146–7
journey maps, 174, 210
joys, 131

K

Kahneman, 170, 175
Keiningham, Tim, 62
KPIs, 222, 236

L

Lean, 226, 245n24
Leela Hotel, 51, 118, 148, 162
legacy environment, in customer orientation, 225
Lewis, Phil, 247
listen, 131, 132
Lockheed Martin, 192
London Symphony Orchestra, 139
'look and feel of website', 83–4
loss aversion, 170–1, 183, 189, 220
loyalty
    in customer experience, 225, 226, 229, 230, 232, 233
    marketing, 208
    points, 145
LUSH, 41, 56, 64, 222, 241n2
Luton Airport, 141, 166–8, 177
LVE, winners of the UK CEM Awards 2015, 137

M

machine data, 209
Mahajan, Gautam, 247
Maklan, Stan, 220, 243n17
Mandarin Oriental, 51, 202, 233
Mankell, Henning, 130
marketing analytics, 207–8
marketing budget, 193
market shifts, 228
Marks and Spencer, 55
Marlow, Nigel, 247
Martens, 132
mass customisation, 171
McDonald, 60
McDonald, Emma, 247
Mears, 58
'Memory Assets', 20
Mesh Experience, 138
metrics systems, 208
Metro Bank, 84, 91, 148, 176, 241n2
Meyer and Schwager, 11–12
Middle East operator, 50
mindset
    acting cross-silo, 192
    automation, 202
    be consistent, 194–5
    brand value, 193–4
    budget for actions, 190, 203
    change in, 194
    client relationships, 197–8
    contract relationship, 195–6
    corporate change in, 190–1
    creative equity, 201–202
    in customer experience, 188
    customer experience, understanding of, 190
    customer value, 195
    discipline in marketing, 193
    embed process and governance change, 191
    evangelist network, 200
    good HR team, 199–200
    heavy metrics, low focus on, 198–9
    maintain flexibility, 194–5
    Opex reduction, 189
    path to progression, 187
    process efficiency, 189
    programme team, 192
    promote cocreation, 201
    relationship sales, 201

sales growth, 189
Mintzberg, Henry, 189
mobile phone buying behaviour, 78
mobile signal reliability, 132
Mobile World Congress, Barcelona, 209
money-grubbing behaviour, 166
Moore, Nigel, 21
Moore, Simon, 19, 161, 174, 247
Morgan Sindall onsite delivery, 233
Morrisons
    convenience, core goal, 177
    emotional omission, 177
    emotional vignettes, 176
    holistic design thinking, 178
    memorable experience, brand goal, 177
    reduce confusion, core goal, 178
    reduce crowding, core goal, 178
    UK supermarket, 175
Mozambique, 130
multichannel integration, 242n4
multi-SIM environment, 88

N

narrative metrics, in customer experience (CX) methods, 130–4
narrative research, 139
NatWest Bank, 143, 160
NatWest mortgages, 233
negative emotion, 157
net promoter, 208
'Network as an Experience Platform', 243n14
Network Green–Customer Red problem, 30
neuronal activity, 129

New Scientist magazine, 177
nonconscious response, 140–1
nonlinear journey mapping, 142–3
Norbert, 156
NPS question
    complicated approach, 109, 110
    customer-focused culture, 112
    customer service, 112–13
    improvement of, 109–110
    managing experience, 111–12
    'negative moments', 111
    recommendation score, 111
    static model, 113–12
NPS score, 210, 223, 227, 238, 243n16, 244n21, 244n23
Nudge principle, 47, 101, 172

O

Oatley, Keith, 130
objective dashboard,, 90–2
objective data, 75–6
objective statistical model, 105
ODC Bank, in Singapore, 192
open listening, 135
Opex reduction, 142, 189, 199, 211, 223, 236, 243n11
O2 store, 78
Overbury, 20, 50, 165, 169

P

Peak-end rule, 175
Perfect Delivery value proposition, 197
peripheral clues, 161
personal and memorable, 36–7, 52, 55–7
personalised campaigns, 208

personality-based segmentation schemas, 175
pester power, 132
Phones4U, 78
pie charts, 131
Pike Place Fish Market, 59, 210–11
Pinder, David, 247
Pine, Joe, 9, 11, 45, 55, 60–61, 196, 221, 222, 230, 243n9, 247
Pixar process, 143
plastic bag, 166–7
Polizzi, Alex, 139
poor performer, 244n23
price attributes, 219
price plan, 132
price quality (PQ), 25, 64
proactive loss aversion, 170
process efficiency, 189, 243n11
product and service drives, 52–3
product attributes, 219, 221
Progression of Economic Value, 221
promoter market, 182
prospect theory, 170

Q

Qantas, 148
quality engineering thinking, 224
quantum computing, 227
quantum physics, 129, 130
Quinlan, Tony, 247

R

rational thinking, 154
regression techniques, 104
regret, 174
relationship marketing, 242n5
relationship value, 36
rep grid, 139
resilience, 182
   complex thinking, 115–16
   Experience brand, 100–1
   improvements, 246
   management, 102
resonance, 245–6
   definitions, 99–100
   Experience brand, 100–1
   management, 102
return on innovation (ROI), 240
   'customer experience differentiation', 45
   Efficiency brands, 44–5
   vendor, 3–4
rewards, loyalty, 232
ring-fencing, 196
RnD, 240
Rolls-Royce, 54
Royal Bank of Scotland, 245n24
Royal Mail, 50, 208
Ryanair, 50
Ryan, Paul, 247

S

Sainsbury's food quality, 195
sales growth, 189
SAM. *See* subjective asset management (SAM)
Schmidt, Bernd, 45
selling experience, 242n4
semantic content (the phenomenology of specific emotion words), 155, 159
sensemaking/interpretative approach, 133
'the sense of the experience' theme, 144

sensory impressions, 130
sentiment analysis, 174, 208
service attributes, 245
service blueprint, 143, 149
service delivery (SERVQUAL), 140, 214–15, 221, 223
service design, 242n4
service excellence, 51
service function, customer experience, 219, 221
service management, 214
SERVQUAL (SPQ), 64
shareability, in CX methods, 131
Sharp, Byron, 20, 245
short-term mindset, 226
Six Sigma, 226, 245n24
Skinner, 230
Smith, Shaun, 236, 247
Snowden, Dave, 77, 89, 130, 131, 133, 134, 226, 242n8, 247
social behaviour, 154
social engagement, 228
social media, 137, 139, 174
somatic markers, 170
sorrows, 131
SouthWest Airlines, 220, 224
Spencer, 55
SPQ (service, product quality), 25
staff–customer–company/brand, 230, 231
staff excellence, 136
staff interaction, 137
stakeholder engagement and ideas, 137–8, 145
Starbucks (the Third Place), 41, 57, 62, 64, 127, 137, 139, 141, 202, 224
Stew Leonard, 25
story metrics, 172

subjective asset management (SAM), 37, 39
  commoditisation, 62–3
  customer drives, 62
  customers score NPS, 63
  deeper drives of customers, 64
  emotions drive, 63
  relationships and durable loyalty, 63
subjective data
  direct effect on behaviour, 77
  managers, implications, 76–7
  modulators, 77–9
  price and product features, 79–81
  qualitative–quantitative, 77
subjective data line
  blue dot effect, 104–5
  companies, customer experience, 103
  curvilinear, 97
  customer satisfaction, 99–100
  hygienic experiences, 98
  perception data curve, 97, 98
  resonance and resilience, 99–100
subjective experiences
  Efficiency, 85
  Experience brands, 88–9
  figure–ground analogy, 90–1
  fuzzy composites, 86–7
  intrinsic quality, 92
  'look and feel of website', 83–4
  objectifying subjective data, 85
  objective dashboard, 90–2
  price-based decision, 89–90
  sense of trust, 84
  statistical root-cause model, 84, 86
  subjective dashboards, 92–4
Swiss insurance company, 199

T

tangibility, 241n2
telco, 43, 195
telecommunications operator, 43, 50, 212
Theme Restaurant Disease, 60–1
'the Theory of Everything', 220
TM Forum, 187
total quality management (TQM), 24
touch points
   unit of analysis, 231
TQM approach, 224
traditional business processes, challenges, 40
traditional surveys
   and customer experience view, 124
   decision moment, 126–7
   inflexible closed system, 124
   measuring experience, scaled surveys, 125–6
   statistical processing regimes, 125
   unit of measurement, 123
Trunkis, 148
'trust or care clues', 136
Type 2 companies, 55, 57

U

UK CEM Awards, 202
uncertainty of conditions, in emotions, 154
United Kingdom, 131

V

valence (positive and negative), 155
value creation, 220, 234
value dimensions, 220
value-in-relationship, 234
value-in-use, 234, 243n12, 245
value tolerance, 230
Vargo, 222
Virgin Atlantic, 145, 172
visual memory, 167
voice of the customer, 208

W

Walker, Julie, 26
Wal-Mart, 55
weak measurements, 130
web browsing speed, 129
Wilson, Hugh, 223, 247
Wittgens, Katharina, 247
www.allaboutexperience.co.uk, 133

X

Xerox, 54

Y

Young, Michael, 29, 34, 236–8, 247

Z

Zappos (we care), 20, 41, 56–7, 64–5, 139, 169
zero defects, 2, 10, 21, 222, 226–7